Hormones, Heredity, and Race

Studies in Modern Science, Technology, and the Environment
Edited by Mark A. Largent

The increasing importance of science and over the past 150 years—and with it the increasing social, political, and economic authority vested in scientists and engineers—established both scientific research and technological innovations as vital components of modern culture. Studies in Modern Science, Technology, and the Environment is a collection of books that focuses on humanistic and social science inquiries into the social and political implications of science and technology and their impacts on communities, environments, and cultural movements worldwide.

Matthew N. Eisler, *Overpotential: Fuel Cells, Futurism, and the Making of a Power Panacea*
Mark R. Finlay, *Growing American Rubber: Strategic Plants and the Politics of National Security*
Jill A. Fisher, *Gender and the Science of Difference: Cultural Politics of Contemporary Science and Medicine*
Finn Arne Jørgensen, *More than a Hole in the Wall: The Story of What We Do with Our Bottles and Cans*
Cheryl A. Logan, *Hormones, Heredity, and Race: Spectacular Failure in Interwar Vienna*
Gordon Patterson, *The Mosquito Crusades: A History of the American Anti-Mosquito Movement from the Reed Commission to the First Earth Day*
Thomas Robertson, *The Malthusian Moment: Global Population Growth and the Birth of American Environmentalism*
Mark Solovey, *Shaky Foundations: The Politics–Patronage–Social Science Nexus in Cold War America*
Jeremy Vetter, *Knowing Global Environments: New Historical Perspectives on the Field Sciences*

Hormones, Heredity, and Race

Spectacular Failure in Interwar Vienna

CHERYL A. LOGAN

RUTGERS UNIVERSITY PRESS
NEW BRUNSWICK, NEW JERSEY, AND LONDON

Library of Congress Cataloging-in-Publication Data
Logan, Cheryl A.
Hormones, heredity, and race : spectacular failure in interwar Vienna / Cheryl A. Logan.
p. cm. — (Studies in modern science, technology, and the environment)
Includes bibliographical references and index.
ISBN 978–0–8135–5969–8 (hardcover : alk. paper) — ISBN 978–0–8135–5970–4 (e-book)
Physical anthropology—Austria—Vienna—20th century. 2. Nature and nurture—Austria—Vienna—20th century. 3. Endocrinology—Austria—Vienna—20th century. 4. Heredity—Austria—Vienna—20th century. 5. Inheritance of acquired characters—Austria—Vienna—20th century. 6. Rejuvenation—Austria—Vienna—20th century. 7. Racism in anthropology—Austria—Vienna—20th century. 8. Vienna (Austria)—History—20th century. 9. Kammerer, Paul, 1880–1926. 10. Tandler, Julius, 1869–1936. 11. Steinach, Eugen, 1861–1944. I. Title.
GN50.45.A9L64 2012
599.909436'130904—dc23
2012023496

A British Cataloging-in-Publication record for this book is available from the British Library.

Visit our website: http://rutgerspress.rutgers.edu

Manufactured in the United States of America

To Glenn and Myrtle

CONTENTS

ACKNOWLEDGMENTS

When years ago, as a graduate student, I read Arthur Koestler's *The Case of the Midwife Toad* I was persuaded by its tale of how a well-meaning scientist had been framed and maybe even murdered by his enemies. Today, after a wonderful intellectual journey, I know better than to accept the simplicity of heroes and villains in history. But that knowledge has come with the often unwitting help of many individuals to whom I am greatly indebted. My interest was piqued many years ago by Merle B. Turner. And I could not have realized it without the tolerance and openness of my colleagues in the Departments of Psychology and History at the University of North Carolina Greensboro. For years, they tolerated my wanderings into the history of the life sciences, often with no clear end in sight. I owe special gratitude to the late Gilbert Gottlieb, and to Timothy Johnston, Kenneth Caneva, Ronald Cassell, and Karl Schleunes.

Discussions and debates with many other colleagues have also shaped my thinking and my approach. I would not have become a "real" historian without the caring help of my great friend Sharon Salinger, who always provided moral, academic, and substantive encouragement. I am also grateful for the support of Mitchell Ash and Richard Burian. This project profited enormously from extended discussions with many colleagues, especially including Mitchell Ash, Sander Gliboff, and Veronika Hofer. I am also grateful for valuable input from Margit Berner, Malachi Hacohen, Michael Hubenstorf, Sarah Jansen, Andreas Lixl, Wolfgang Maderthaner, Arndt Niebisch, Klaus Taschwer, Sarah Wagner, and the members of the old UNCG German Studies Group. The chance to share ideas in the classroom with students also clarified my thinking on many, many issues. Colleagues and students have far more impact than they know.

Laura Edwards, George Michel, Joan Paluzzi, Sharon Salinger, and Karl Schleunes provided extremely constructive comments on earlier drafts of several chapters. They gave me valuable time and offered great suggestions. Generous advice on numerous translations from the original German was patiently provided by Andreas Lixl, Arndt Niebisch, and Susanne Rinner; and Paul Silvia and the UNCG Agraphia group kept me going amid moments of polygraphic confusion. I am grateful to all. Thanks are also due to an anonymous reviewer for excellent suggestions on the book's concepts and structure and to Marsha Richmond for an extremely helpful reading of the entire work. My editor, Peter Mikulas, at Rutgers University Press has been a boon throughout. He offered clear suggestions for revision at various points, saw me through several problems, offered terrific encouragement, and in the process greatly improved the final work. And Margaret Case provided especially careful help in polishing the final product. My gratitude to all, of course, implies no responsibility for any remaining errors. All are of my own making.

Financial support for the project was provided by grants from the National Science Foundation, the Fulbright Scholars Program, the UNCG Research Assignments Program, the UNCG International Programs Center, and the UNCG Kohler Fund. They made it possible to conduct research at archives in Austria and the United States. The staff and administration of several libraries and archives provided important information that I might never have found without their help. I am especially grateful to Roy Goodman and Valerie-Anne Lutz at the American Philosophical Society, Liana Zhou and Shawn Wilson at the Kinsey Institute, Karl Sablik and Manfred Skopec at the Institute for the History of Medicine at the Medical University of Vienna, and Arlene Shaner at the New York Academy of Medicine. The staffs of the Vienna Rathaus Library, the Rockefeller Archives, the Austrian National Library, and the University of Wisconsin Special Collections Department were also very helpful.

Finally, I have been truly honored by the hospitality shown me by the William Tandler family and the Hugh Iltis family. Their generosity and insights were a special treat that helped me see "figures" as real people. They welcomed me into their homes and shared important documents and memorabilia as well as their lovely memories of Tandler, Kammerer, and Iltis.

Hormones, Heredity, and Race

1

Three Failed Scientists

Nature and nurture were pitted against one another for most of the twentieth century. In the minds of many, they still are. Is an attribute learned or instinctive? Or is it a mix, more learned or far more hereditary? This kind of separated partitioning was, however, not prominent at the beginning of the century. At that time, many important thinkers still accepted the long-held idea that persistent changes in the environment could alter the nature of heredity, so that an individual developed under the influence of heredity that was molded by his or her environment. There could even be several kinds of heredity, some of which were especially amenable to environmental change. These thinkers fused heredity and the environment in a way that made the question of which contributed more to a trait either irrelevant or meaningless.

Scientific debates on the issue began in earnest in the 1880s with the ideas of the German zoologist August Weismann and the rise of the new genetics. These debates turned on one of the most contentious concepts in the study of heredity in the early twentieth century: the issue of the inheritance of acquired characteristics. Can offspring inherit *as hereditary* attributes that their parents have acquired? Weismann denied it. This book tells the story of three Austrian scientists who challenged Weismann and, in the years around World War I, proposed a modern explanation of how the inheritance of acquired characteristics might occur. They believed that hormones were the key. Through hormones, individual adaptations—acquired adjustments that improved life in one generation—could, once entrenched via exercise or practice, become hereditary in subsequent generations. The sex hormones, the three men argued, were especially likely to mediate these environmental changes in heredity.

Weismann had proposed that the hereditary material (then called the "germ-plasm") was isolated and sequestered. It lay hidden deep within "germ" cells, and it was transmitted continuously from parent to offspring in ways that made it immune from environmental influences within the body. The implications of his proposal extended beyond the science of heredity to touch on social debates about the nature of organic progress. Was nature random, or did it contain some principle that ensured forward-moving progressive change in life? Weismann and the British philosopher Herbert Spencer debated the issue in the 1890s, and blind versus progressive views of organic change informed their different views of heredity. Spencer defended the inheritance of acquired characteristics as a force shaping human purpose. Weismann sought accounts of change that relied only on chance and natural selection, Charles Darwin's new account of evolution. Both men drew on evidence to explain odd biological phenomena that, on the surface, seemed to have little to do with purpose or progress—such things as cave dwelling animals, whose eyes degenerated when they lived in totally dark environments. Weismann insisted that natural selection alone could explain them; Spencer argued that they stemmed from the disuse brought by the dark cave environments, disuse that eventually made them hereditary.

Weismann did experiments removing rats' tails to demonstrate that environmental change was ineffective. But he was on weak ground, arguing a point based on something that didn't happen. As he acknowledged, his conclusions were drawn more from first principles than from evidence. In 1893, he wrote: "we know indeed, nothing at all but the chief foundation of the process; and therefore anyone who does not comprehend the logical necessity of the theory, or will not recognize it, can easily set aside the individual instances as untenable."[1] Even by 1920, evidence had not decided the issue. And by then, the stakes were especially high. Many believed that humanity itself was deteriorating; if there were no purpose in biology, what ensured nature's progress? If nature contained principles of inherent purpose, humanity would resolve its problems naturally; nature would take care of itself and of human beings. If it did not, humans would have to step in to intervene to direct their own futures. But who would decide what that direction would be? And what form would the intervention take?

The scientists on whom I focus believed that they had a novel solution. They agreed, as most did, that the environment could be destructive. But without accepting notions of inherent purpose, they also argued for the constructive power of the environment as the key to the problems of heredity and progress. The environment could, after all, shape adaptive change in heredity through the intermediary action of the body's hormones, chemicals produced by glands that secreted their products directly into the bloodstream. It was a novel hypothesis, and it raised intriguing possibilities. But they failed, and they failed spectacularly.

Modern science can be and has been of huge benefit to humanity. The histories of medicine, the cell theory, and the germ theory of disease, for example, reveal many important successes. But scientists have also pursued flawed ideas. Some of these are immoral; some are unsupported by evidence; some are just discarded. And some seem wrong, but can return, usually in very different dress, to be reconsidered at a later date. Historical accounts of failed science are not stories of the linear accumulation of flawed knowledge, which at some point reaches a logical criterion established for failure. The information is not usually discovered to be wrong because evidence tells us that it is. That happens; but perhaps just as common are episodes in which ideas are discarded because they are premature or inconsistent with a social or political context whose solution for a time drives scientific priorities. Scientists and the problems they face cannot be separated from their social contexts. Personal, political, and social influences shape scientific communities and the motivations and beliefs that guide scientists' work. They help shape the "configuration of ideas" that makes science a useful and socially meaningful enterprise, and, at some points in history, they influence how the rules of science work.[2]

Perhaps no problem illustrates this more clearly than the early twentieth-century debates about nature and nurture that informed the rise of eugenics. Enlightenment ideals of human improvement, Darwinism, the rediscovery of Mendel's laws of heredity, and the growth of modern genetics all supported the new field of eugenics, a branch of applied biology that sought to intervene and modify heredity to improve humanity. Many eugenicists rationalized numerous ethical abuses in the name of science and progress. These including forced sterilization, euthanasia, and even the murder of people deemed "unfit." Those advocating and implementing eugenics were not, however, primarily fanatical politicians or pseudoscientists lacking a solid grounding in science (though there were some of these). Instead, many were well versed in the expert scientific knowledge of the time and deeply involved in scientific debates that were decided—as many are—with incomplete evidence. But there were also challengers. In the 1920s, several dissenters challenged the ascent of mainstream eugenics, which I call selectionist eugenics. Some dissenters resisted eugenics altogether, basing their rejection on moral, not scientific, grounds. Others resisted the science underlying eugenics. Some did both.

Dissenting Scientists

The three scientists that I present were among the dissenters. In the interwar period they promoted an alternative application of biology to society based in their hormonal interpretation of the inheritance of acquired characteristics. Genetics and endocrinology were both new sciences, and these men were at the center of an attempt to integrate the two fields to redefine how humanity might ensure its future. They lived in Vienna, and they knew each other distantly. Two were coworkers. One was an acknowledged pioneer in his field, a second was a prominent scientist-politician, internationally known for modern health reforms. For a time the international scientific establishment paid close attention to their ideas; and for a time their approach succeeded. But when it failed, it failed amid racist ridicule, political exile, fraud, and even suicide. Theirs is the story of how controversial but good science, science judged plausible by the standards of the time, can fail partly under the weight of political and social influence, personality, and prejudice. It is the story of how scientific failure can be abetted by the historical contexts in which scientists live and work. I tell their story through their personal passions and the overlapping concepts that united their effort. They had different goals, and each brought distinct expertise to the task; but they shared a deep regard for developmental flexibility understood as a foundational dimension of biology. And they were committed to building biological principles that resisted the prevailing social hierarchies of race, class, and gender at just the time when these hierarchies had begun to dominate mainstream eugenics. Through their history we can understand how dissenting voices in science—in this case Jewish dissenting voices raised in a time of virulent antisemitism—fail as much for political as for scientific reasons, and with spectacular personal and social consequences.

One of the three was a zoologist, one an endocrinologist, and one an anatomist. Two were internationally famous; the third was known, but less prominent. The least prominent man was the zoologist Paul Kammerer (1880–1926). For Kammerer, the failure was both scientific and deeply personal. In 1923, while on lecture tour in America, he had been over-enthusiastically hailed as "Darwin's Successor" by the *New York Times*. But late in 1926, he committed suicide in the wake of intimations of scientific fraud when the journal *Nature* announced that some of his experimental specimens had been artificially altered. The most prominent of the three men was the pioneering reproductive endocrinologist Eugen Steinach (1861–1944). In 1926, Steinach was among the most famous endocrine physiologists of Europe. His failure was scientific and medical. By the 1940s, he was ridiculed and his scientific reputation destroyed. He was on vacation in Switzerland when the Nazis entered Vienna in 1938; he was exiled and died there, a bitter and unhappy man. The third was the anatomist and politician Julius Tandler (1869–1936). Tandler's failure was scientific and deeply political. After the end of World War I, he became a central figure in Austrian public health. He shaped the health and child care reforms of Vienna's governing Social Democrats, spearheading the medical and health reforms of "Red Vienna." But after over fifteen years of civil service, in 1934 he was driven into exile under the threat of imprisonment. Two years later he died of a heart attack in Moscow.

The three were colleagues in the Viennese scientific community. Steinach and Kammerer worked in the same laboratory. Steinach and Tandler had been students in the same institute; both were experts in the science of sexuality, which greatly benefited from their pioneering work. Kammerer and Tandler were committed to the humanitarian reform of society on scientific and socialist grounds. All three worked on an application of biology to society that united endocrinology and genetics to emphasize developmental flexibility over and above rigid genetic fixity. And for a time, they explicitly argued from within biology against the eugenics that advocated the forced sterilization of mentally and physically disabled people and the elimination of supposed "degenerates." In the development of scientific knowledge, social bias can insert itself into spaces left empty by incomplete theories and evidence. Amid the social structures that shape and apply scientific knowledge, scientific advance is often shaped by such bias. The reactions within science to these men's views on sexual equality, to their political positions, to their ideas of racial flexibility and improvement, and to their perceived Jewishness, eventually helped engulf their science.

The least significant in his own time was Kammerer; but partly because his passion linked the other two around a novel solution to the nature-nurture problem, I begin with him.

Paul Kammerer

If Kammerer is remembered at all today, it is probably for the cloud of fraud recounted by Arthur Koestler in his 1971 book *The Case of the Midwife Toad.* The book describes an episode in which Kammerer was implicated in tampering with data that he had collected around 1909. The data examined the mating characteristics of land-breeding toads that had been reared in water, the so-called midwife toads. From about 1910 to the discovery of fraud in 1926, the data were important because (along with many others) they demonstrated experimentally that characteristics acquired by individuals in one generation, through positive adjustment to a changed environment, could alter the heredity of their descendents—the inheritance of acquired characteristics. The idea had been popular in the early and middle nineteenth century, but by 1920, its fortunes had begun to decline. Against the rising tide of the new science of genetics, Kammerer remained one of its most vocal defenders.[3]

But Kammerer was a poorly paid assistant who had labored for many years over experiments on toads and salamanders. Such a journeyman scientist has an important place in this story for two reasons. For almost two decades he had amassed a large body of zoological evidence considered by many to be the most convincing data demonstrating the inheritance of acquired characteristics. By 1920, if the idea were to stand or fall, many believed it would do so based in large part on Kammerer's evidence. Second, beginning in about 1918, Kammerer developed a hypothesis that drew on a new theory in reproductive endocrinology to argue that hormones could mediate the persistent and flexible impact of the environment on heredity. In hormones, he sought evidence of the mechanism by which heredity might be rendered flexible. Kammerer and his senior colleague, endocrinologist Eugen Steinach, then collaborated on experiments analyzing the impact of climate on heritable differences in the sexual characteristics of mammals. The results led them to reject what were thought to be rigid sexual differences among human "races." Kammerer's goal was a scientifically based moral reform of what he saw as the racist base of mainstream eugenics. He advocated instead a new, more ethical eugenics, which he called "productive eugenics."

Many scientists carefully watched Kammerer's research. In 1924, American zoologist and geneticist Herbert Spencer Jennings wrote to his colleague, Yale embryologist Ross Harrison, endorsing Kammerer's work. Jennings had just hosted Kammerer's lecture at Johns Hopkins University, and after the lecture, Kammerer stayed in Jennings's home. It is not clear why Jennings felt compelled to write to Harrison, but the communication reads like a letter of recommendation. In it, he gave Kammerer measured praise. He was, wrote Jennings, adept at raising animals and probably a "very skillful experimenter." Jennings also commented on Kammerer's personality: "He also seemed to me a very direct, unaffected, straightforward personality; certainly nothing about him suggests the faker"—confirming that others were already suspicious of his research. But Jennings was no sycophant; he had no reason to be. He warned that Kammerer might not be good at theory; he had a "simplicity about him. . . . On the other hand I should be disposed to have confidence in any particular concrete result that he presents." At least some of Jennings's Hopkins colleagues apparently agreed: "The other men who met him seemed to get the same impression that I did, so far as I heard, as to his personality and honesty." Alluding to controversies that already surrounded Kammerer's work, Jennings criticized the "bad tactics" of Kammerer's American detractors.[4]

Jennings was well informed about Kammerer's research. In lectures presented before and after Kammerer's visit, Jennings examined in detail Kammerer's zoological evidence for the inheritance of acquired characteristics. The debates had intensified around the experimental evidence supporting it, and American scientists were particularly skeptical. Jennings's lecture notes puzzled at the strong reaction against the concept in America. Any serious consideration of the idea, he wrote jokingly, quickly convinced one's scientific colleagues that he was probably guilty of insurance fraud or of beating his wife. Jennings noted that the positive evidence for the inheritance of acquired characteristics would have to be overwhelming. But he considered Kammerer's work promising—the best evidence available. After describing Kammerer's salamander studies in detail, he wrote "remarkable if true!" Of Kammerer's body of work, he noted: "These 14 or more years of work gave [a] very great mass of results of very highest interest—published in extensive detailed illustrated memoirs in the scientific journals on many species."[5]

His lectures were a strong endorsement: "If work proves itself,—[it] will undoubtedly have to be considered one of the most important pieces of work in generations. . . . Important to keep eyes on Kammerer's experiments; no one will catch up with him for a while!"[6] But this was almost as far from prophetic as one could get; less than a decade later Jennings wrote: "[Kammerer's] work made a strong impression, until it was suddenly shown that the result of one of his experiments had been falsified. Kammerer committed suicide. His entire work has been discredited."[7] After the scandal broke in the summer of 1926, Kammerer denied any role in the fraud, and many colleagues defended him. But, already a depressed man and now dejected at the damage to his honor, he shot himself in September. As Jennings shows, the suicide was widely regarded as an admission of guilt. Kammerer's reputation was destroyed, and with it, the most convincing evidence for the inheritance of acquired characteristics.[8]

Eugen Steinach

Paul Kammerer was a master curator of animals. Breeding and rearing healthy animals were his passions, and they had enabled him to produce convincing zoological evidence. But after 1900, knowledge of cell structure grew rapidly, and the requirements of evidence changed. Zoological demonstrations of flexible heredity like Kammerer's were less and less convincing. Instead, cellular evidence was needed to explain *how* flexible heredity worked. Scientists wanted to know what cellular process could convey an environmental change to deep within a cell and alter its hereditary material. In about 1917, Kammerer abandoned his salamander and toad work and turned to the research of his new colleague, Steinach, to address the problem of mechanism.

Steinach had worked for years with the renowned physiologist Ewald Hering at the German University in Prague. But in 1912, he was hired to direct the Physiology Department in Vienna's Institute for Experimental Biology, where Kammerer had worked since 1903. Though his early work was on the nervous system, by the late 1890s, Steinach had turned to the study of what were then called inner secretions—hormones. Between 1910 and 1930, he carried out ingenious experiments showing that sexual development in mammals was not rigidly controlled either by genes or by gametes (eggs and sperm). Rather, sexual attributes unfolded flexibly under the influence of hormones. He also demonstrated that hormones were produced in a bisexual mix that could shape the development of either male or female sexual characteristics in either males or females. His work challenged the rigidity, if not the form, of the late nineteenth-century boundaries between male and female, and his scientific demonstrations were often sensational. The program established Steinach as the ranking, if controversial, pioneer in Central European reproductive endocrinology. But Steinach's hopes were also based in endocrine procedures that he had developed to treat the infirmities of old age by surgically modifying hormone production. The method, termed rejuvenation, was a surgical forerunner of hormone replacement therapy, and it may well have been his scientific undoing. By the late 1930s, rejuvenation was ridiculed and Steinach's pioneering early research marginalized.[9]

Julius Tandler

Though Kammerer's suicide was tragic, his failure was spectacular not because it was linked to scientific fraud or because it ended in self-inflicted death. It was spectacular because in post–World War I Austria, the hope of a different kind of application of biology to society rested with the validity of the inheritance of acquired characteristics. Kammerer had long advocated the concept for improving society, and he was not alone. Many Viennese scientists and social theorists based their hopes for an improved human heredity on the inheritance of acquired characteristics. In Vienna by 1924, the year Jennings endorsed Kammerer, Julius Tandler was arguably the most powerful of these hopeful reformers. Tandler was a converted Moravian Jew who had risen to unusual prominence at the University of Vienna. In 1910, he became professor and director of the Anatomical Institute; by 1914, he was dean of the medical faculty.[10]

Tandler was an expert on the anatomy and physiology of sex. Near the end of World War I, he moved into politics to apply his scientific expertise to aid in restoring the health and welfare of the Austrian people in the wake of the war's devastation. In 1920, he became municipal councilor for welfare (*Wohlfahrtspflege*) in the socialist government of Vienna; and by the mid-1920s, he had implemented reforms that were intended to reinvigorate the war-torn, hungry, and tuberculosis-ridden society. Under his leadership, even its most downtrodden citizens would have a place to live, some food, and a lower risk of disease. Tandler showed special concern for Vienna's children. On reporting his arrest in Vienna during the political upheavals of 1934, the *New York Times* praised his reforms and wrote, "his work in child hygiene has found particular recognition."[11]

Tandler's science was consistent with the socialist philosophy of postwar Vienna, in part because it rested on the inheritance of acquired characteristics. But his significance stems from the link between his scientific work and his political influence. He implemented the vision of Vienna's Social Democrats, which held that rationally restructured public housing, hospitals, children's education, and health were the rights of all citizens. Tandler, like Kammerer, believed that such rights could, through the inheritance of acquired characteristics, transform humanity biologically. But the cloud of fraud surrounding Kammerer's suicide jeopardized the scientific underpinnings of this vision. How did Tandler's policies change in the difficult years after the suicide—years of tense political turmoil? Did Tandler, like Jennings, view the suicide as an admission of guilt and abandon the concept that empowered his reforms? Or did he continue to defend the flexibility of heredity and its implications for human improvement? I suggest the latter. Despite much linguistic hedging, as late as the early 1930s Tandler continued to endorse and implement an approach to eugenics understood in terms of hormones and flexible heredity.

Scientific Failure and the Inheritance of Acquired Characteristics

The concept of failure has long been used to understand fin-de-siècle Viennese culture. But rarely does it touch Viennese science. Quite the contrary, as the artistic and literary generation supposedly rebelled against the rationality of its parents and abandoned liberalism to retreat into internal reflection and aestheticism, Viennese science flourished.[12] The University of Vienna's medical school was renowned at the time; Viennese traditions in physics and philosophy were groundbreaking: they led to the huge impact of the Vienna Circle on Western science and to important connections between science and social reform. But scientists always debate and challenge; a dialectic of evidence, competing schools, and research traditions ensures that as one side prevails, the other (or others) fails. In one sense, failure is fundamental to the logic of falsifiability, the Viennese principle that holds that hypotheses must be meaningfully refutable or they are not really scientific at all. It requires that all scientific concepts at least imply the potential to be wrong. But that is different from the abandonment of concepts that are rejected as either categorically wrong or simply shelved. And, as examples such as continental drift show, discarded ideas can return even to dominate.[13] Such a return is occurring in our current understanding of the relationship between genes and environment. Examining how rejected concepts fail, only to become prominent again in much different dress, can reveal much about the process of failure and return, as competing paradigms challenge one another.

I propose that as the scientific tide begins to turn toward new and intriguing ideas, success becomes self-reinforcing. In doing so, it generates an asymmetry between success and failure. Successful paradigms become even more right, while their competitors remain just wrong enough to be ignored. Successful research traditions and the communities that shape them grow disproportionately, while their competitors disappear. One result, as success feeds itself but failure does not, is that scientists are able to say much more about why one idea is right than about why another is wrong. That relative rightness, not formal disconfirmation, makes the failing approach "wrong," so that opposing ideas and evidence are quickly rejected. With more asymmetry, success feeds even more success. This accelerating momentum of success can have many sources. Convincing empirical work is surely a part of the self-reinforcing process. The commercialization of scientific evidence (just becoming prominent as this story takes place) is, for example, part of the engine of success. Commercially available equipment and animals privilege particular methodological solutions and accelerate their value for certain problems. So do the social structure and language of science: the emergence of new disciplines represented by journals controlled by the winning side; the founding of institutes promoting one side's assumptions; and the development of new terminology understandable only under one set of assumptions—all feed success.

But beyond the empirical, structural, and conceptual efforts that drive science internally, sociocultural influences and timing also contribute to the asymmetry of success and failure. Kammerer's, Tandler's, and Steinach's ideas were immersed in social questions about sex, gender, and race at a time when the significance of each was hotly debated in Euro-American society. The early study of sex hormones occurred in a sociocultural context that altered the scientific stage, and the relevance of these scientists' ideas to social considerations of race and gender—to antisemitism, to attitudes toward people of color, and to the "woman question"—shaped their thinking and blunted the power of their ideas.[14] When scientific change overlaps in time with powerful political trends, dissenting voices that are aligned with opposing political positions will have less and less impact. As the momentum of success grows disproportionately on the other side, external influences can gain reciprocal support by a kind of snowballing of oversized scientific success.

The fate of Kammerer, Tandler, and Steinach's endocrine hypothesis illustrates the power embedded in the engine of success. After the turn of the century, that engine was the nuclear chromosomal model of heredity bolstered by Weismann's assumptions of an isolated germ-plasm resistant to environmental adaptation. Following the rediscovery of Mendel's laws of heredity and the development in America of *Drosophila* genetics, modern genetics accelerated partly as a result of its own success. By 1930, there was increasingly little room for innovative alternatives to gain an explanatory foothold and argue how the environment might alter heredity to instill adaptive traits. The story of Kammerer's, Tandler's, and Steinach's failed attempts highlights this intriguing process. I do not claim to solve the problem of failure. Rather, I present the attempt to integrate endocrinology into a mechanistic account of the inheritance of acquired characteristics as a plausible, but failed, effort to render heredity flexible during a turbulent period, an effort that was discarded in the wake of the growing success of twentieth-century genetics. In the process, a core of scientific resistance to the rise of Hitler's Germany was silenced, with the disturbing result that one of the great scientific advances of the twentieth century (modern genetics) overlapped in time with one of the great evils of the twentieth century.[15] Examining this episode on the eve of that convergence can focus our attention on those features that shifted the momentum of alterative paradigms to trigger the disproportionately greater success of the favored.

Conceptual Themes

Three problems emerge from a nuanced historical treatment of the inheritance of acquired characteristics in this period. The first reflects the odd fact that in the late nineteenth and early twentieth centuries, when the concept of the inheritance of acquired characteristics was still prominent, the effects of the environment were treated asymmetrically. Under the influence of the notion of biological degeneration (Chapter 2), damaging influences of the environment on heredity were stressed over and above any recognition of the positive impact of the environment. Many thinkers who accepted hereditary damage wrought by environmental influence vigorously denied that any beneficial change was possible. Kammerer, Tandler, and Steinach challenged this unevenness to present a constructive view of hereditary change, shaped by the environment and understood in terms of biological development and "direct" adaptation.

The second involves the uses to which their endocrine hypothesis was put, especially in debates about eugenics. Kammerer's and Tandler's hypothesis drew on the distinction between somatic induction and parallel induction—a distinction introduced by Weismann, and which by 1904 framed the terms of the debate (Chapter 3). By proposing internal processes by which the soma could act on germ to induce adaptations, they framed somatic induction as a productive process that could act just as readily to mold beneficial as damaging change. This enabled Kammerer, Tandler, and Steinach to argue for the environment as a force for improvement, for regeneration, and not just for hereditary degeneration (Chapters 2, 5, and 7). As such, they countered the tide of degenerationism that, after World War I, fed the link between racism and eugenics. Hereditary improvement via environmental reform became the basis for Kammerer's productive eugenics (Chapter 5) and for Tandler's many municipal reforms (Chapters 6 and 7).

Finally, Kammerer explicitly challenged the racist base of mainstream eugenics, which assumed the hereditary inferiority of those deemed racially or medically "unfit." He argued instead that environmentally modified and hormonally transformed improvements in heredity could eliminate biological differences among human "races" and help restore health and morality. But Kammerer, Tandler, and Steinach were each of Jewish heritage. And they pursued their ideas during a period of virulent racial antisemitism, when a challenge to the race concept could be interpreted as a flaw in the "race" of those mounting the challenge. In this context, antisemitism played a significant role in the scientific and social reception of their ideas (Chapter 8). These three scientists failed in part because they were Jewish.

Though both drew on Steinach's pioneering science, Kammerer and Tandler are more central to my story than Steinach. My treatment of their impact is built around two theses. First, Kammerer's work must be seen in the context of his commitment to fighting the racism that he saw growing in selectionist eugenics. It was this commitment that motivated his risky switch from zoology to endocrinology to seek a mechanism for somatic induction. Kammerer's political and social involvement was intimately connected to his science. It embodied the epistemological, ethical, and political views that motivated his approach to heredity and his endocrine attempt to reform eugenics after 1918. Second, though his approach was more scientifically cautious and his efforts more politically circumspect than Kammerer's, Julius Tandler built the foundation of his welfare reforms around the ethos of developmental flexibility that he shared with Kammerer, and around their shared idea that hormonal mechanisms could explain the constructive impact of the environment on heredity. These ideas had been the basis of his scientific work, and through the 1920s he continued to accept the idea that environmental reforms could, via the endocrine system, improve human heredity. Despite Kammerer's many impassioned statements of the positive impact of somatic induction, Tandler never explicitly referred to Kammerer's productive eugenics. I argue that after Kammerer's death in 1926, Tandler nonetheless continued to defend the spirit of their common endocrine approach to productive eugenics.

A number of other relationships emerge from their story. For example, a philosophical residue of teleology (explanation in terms of goals and purpose) plagued the idea of the inheritance of acquired characteristics. The reaction to Kammerer's research, which rejected inherent purpose in biological processes, shows how some of those defending mechanistic causation over teleology in biology did so by accepting a rigid hereditary determinism (Chapters 3 and 9). Similarly, the specific gonadal cells proposed by Tandler and Kammerer to be the keys to their endocrine solution, the interstitial cells, were embedded in intense social debates about the biology of human sex differences (Chapters 2 and 4). Here too, social issues inserted themselves into scientific deliberation. In several ways, therefore, examining the context surrounding these scientists' attempts to develop an endocrine solution to the problem of flexible heredity reveals a tangle of social and political influences that altered the scientific debate and facilitated their failure.

Converging and Diverging Lives

These three men lived quite independent lives. But their science and their social missions converged during three periods in the early twentieth century. The first occurred between 1910 and 1914. This was when reproductive endocrinology emerged as a laboratory science in which both Tandler and Steinach were pioneers. In 1910, the first German-speaking textbook of endocrinology was published; at about the same time, Steinach returned to Vienna and turned his attention from neurophysiology to reproductive endocrinology. In 1913, Tandler's widely known reputation in sexual anatomy extended to sex hormones with the publication of his award-winning monograph, *Die biologischen Grundlagen der sekundären Geschlechtscharaktere* (The biological foundations of the secondary sexual characteristics), and with his proposal that a new theory about the sources of sex hormones might also be the key to flexible heredity.[16]

The second was a period of convergence between 1918 and 1926. The end of World War I and the fall of the Austro-Hungarian Empire brought enormous change in Austria and in Vienna. Politically, the most striking change was the socialist revolution that "transformed" a patriarchal monarchy into a socialist republic. But by 1920, the Austrian socialists had retreated to Vienna, and Tandler began the political career that immersed him in the reforms needed restore the health of the Viennese citizens. At the roughly same time, Kammerer strengthened the scientific basis of the inheritance of acquired characteristics, linking it to Steinach's endocrinology. Their collaboration (Chapter 4) confirmed empirically the power of sex hormones in altering heredity and supported Kammerer's revision of the biological bases of racial differences in human sexuality. It was the key scientific piece in his rejection of fixed racial hierarchies, and it strengthened the foundation of his productive eugenics. For all three men, this was a time of hope. Their science was expanding, and their prospects and vision were strengthened by Tandler's powerful position in Vienna's municipal government.

The third period entails an odd convergence because it began with Kammerer's death. This was the period of decline that followed Kammerer's suicide in 1926 and lasted until the Social Democratic Party was declared illegal and its leaders imprisoned in the Austrian Civil War of 1934. After 1926, political tensions increasingly pitted socialism against entrenched antisemitism, the emergence of fascism, and calls for racial approaches in Austrian eugenics. Each trend complicated Tandler's task. Numerous other changes occurred in science and society that facilitated the decline of their ideas. The chromosomal theory of heredity based in Weismann's assumptions made great strides; and for many, those advances provided support for the compulsory sterilization programs of selectionist eugenics that, by this time, were flourishing in America. By the end of this period, Central European race science had grown so significantly that the work of three scientists of Jewish ancestry was suspect on racial grounds alone.

In none of these periods were the three men a team, nor did they ever work as one. They had different goals, distinct personalities, and they followed independent paths. Despite the shared foundation of their idea, they were rarely in agreement—socially, politically, or personally—about every aspect of science or its application to society. But the connections among them, their agreements and disagreements, developed amid the storm of discussions about sexual science, heredity, eugenics, and race that engaged Central European intellectuals in the first third of the twentieth century. In pursuing progressive reforms initiated by Vienna's Social Democrats, they shared a growing respect for individual human variety, a naturalistic view of society improved by life science, and a deep appreciation for biological flexibility and development, which each understood in terms of the endocrine system.

This book is not a biographical sketch of each man; nor is it an attempt to provide comprehensive histories of endocrinology, eugenics, or genetics. Rather, it is a triangulation in which each man provides a key piece of an interlocking scientific story in which hope and political change for at time offered an ethical alternative to the promise of improving humanity through the application of biology to society. Eventually, the efforts of these scientists were taken up by powerful eddies in the social currents that shaped debates on one of the central problems of biology in the early twentieth century: the nature-nurture problem. Their intersection, the intersection of their science, their personalities, and their ancestry, played central roles in their hope for a renewed humanity based in a humane eugenics and in its downfall. As their failures unfolded, the scientific and political base of an alternative eugenics was silenced. Selectionist eugenics had already swept over America; and even more tragically, it eventually swept over Central Europe. The attempt to understand somatic induction via reproductive endocrinology to save the inheritance of acquired characteristics reveals more than the history of a failed attempt to unite endocrinology and heredity. It shows that plausible science sustained by reasonable argument can be overwhelmed by politics, personality, social prejudice, and the over large momentum of newfound scientific success.

PART I

Constructing Heredity

2

Rehabilitating Sexuality

Degeneration versus Development

Richard von Krafft-Ebing was the acknowledged pioneer in the late nineteenth-century study of alternative sexualities, then considered deviant pathologies, and he was near the center of the Euro-American obsession with medicalizing sexual diversity in the early twentieth century. If, the thinking went, pathological sexual deviations from "the normal" could be classified, diagnosed, and defined, they could also be managed. And management according to prevailing social standards would preserve sexual restraint, protect the healthy norm, and maintain a sexual division of labor by which men would control society and women would prevail in the home. Krafft-Ebing was known for his extensive use of personal case histories, many of which led to his authoritative listing of over thirty different types of "perversions" that were believed to threaten the prevailing social order. Among them were anesthesia sexualis, epilepsy-perversion, public indecency, fetishism, necrophilia, and masochism—all terms that Krafft-Ebing had coined.

In the late 1880s, a psychiatric patient who suffered from what at the time was called "pederasteric psychic hermaphroditism" wrote Krafft-Ebing and movingly described his inability to love women:

> My nature is absolutely hermaphroditic in a psychosexual sense. (Whereas I can only sensually desire women, I can love as well as sensually desire youths). If marriage between men would be possible, I think I would not avoid a lifelong union, while marriage with a woman seems to me to be something impossible. . . . A constant association with a youth who is physically pleasing and in mental harmony with me, who could understand all my feelings and share my intellectual opinions and endeavors, would, it seems to me, be the greatest happiness.[1]

Krafft-Ebing viewed his correspondent's affections and his sexual preferences as signs of disease, because "the greatest happiness" would come only with other men. His inclinations reflected not a moral lapse that had resulted in questionable conduct, but a disease that pervaded his constitution and probably could not be treated. The disease had destroyed his will, and, much as he might be unable to will away the bacteria causing tuberculosis, he was powerless to act or feel differently.

Krafft-Ebing's approach, like many others, was dictated by the powerful degenerationist paradigm that dominated biology, medicine, and psychiatry in the late nineteenth century. That paradigm equated disease with biological degeneration and tainted heredity. Patients often accepted and incorporated this discourse, even berating themselves for their fates. And as the above case shows, their stories often flew in the face of the ideal of romantic love, viewed then as a biological possibility only between a man and a woman. Krafft-Ebing's humane response to this patient and to others who wrote him with their stories was to see their "perversion" as an illness for which the individual could not be held responsible—one that demanded treatment, not incarceration, which was still the fate of many. The patient's constitution was permanently impaired in a medical fusion of somatic, neural, psychic, and moral deterioration for which he could not be held responsible.

Medical degenerationism emerged again in the 1930s to justify Nazi atrocities in Central Europe. But the scientists who promoted it by that time did so in clear view of other, more humane alternatives. For in the 1920s, degenerationism in biology was challenged by a different paradigm of sexuality. This approach stressed flexibility in sexual development and the continuity between "the normal" and various departures from the norm. Reproductive endocrinologist Eugen Steinach helped shape this new paradigm. He and others viewed sexual diversity, not as biological degeneration indicative of inner deterioration but rather as the product of hormonal action that was qualitatively no different in individuals considered "normal" and those deemed "degenerate." Natural chemicals found in all people could be altered during development to produce altered sexual outcomes. And in the minds of some sexual reformers, this meant that variant expressions of sexual behavior and desire, such as homosexuality, were the product of the same "natural" processes in "normal" and in "deviant" individuals.

Steinach's position was based on a controversial new theory in reproductive endocrinology. The theory, first proposed in 1903, focused on an innocuous type of cell discovered in the testes. The cells, termed interstitial cells, became the basis for a radical revision in the biology of sexuality, one that distinguished between the purely reproductive functions that give rise to offspring and the sexual functions that shape psyche and behavior, but might have little to do with reproduction. Challenging degenerationism, which was especially entrenched in debates about sexuality, this view became the basis of a prominent post-war Viennese tradition in reproductive endocrinology. The scientific and social debates that surrounded it touched on sexual diversity, on the "bisexuality" of men and women, on homosexuality, and on human hermaphroditism, scientific questions that could not be separated from the perceived threats that fed the fears of middle-class degeneration.

The Background of Degeneration in Science

In the late nineteenth and early twentieth centuries, European and American biology often explained the social and psychological realms reductionistically, as manifestations of biological processes. The biological sciences grew enormously in this period, and the new fields of neurology and psychiatry offered authoritative explanations of human beings, their minds, and their conflicts. Things that people had done or been for centuries were, by 1890, classified as diseases, as abnormalities. In the process, many expressions of human variety were labeled and condemned.[2] Krafft-Ebing pioneered the labeling process in psychiatry, and he illustrates the mix of degenerationism, medical humanitarianism, cultural bias, and morality that dominated Central European society at the turn of the twentieth century. Science, politics, morality, and disease were of a piece.

That piece was held together and strengthened by two broad trends that complemented one another in European science and society from the mid 1850s through World War I: the fear of degeneration and the hope for progress. Progress and improvability were the social heirs of the Enlightenment's hope for a better life through naturalism, reason, and science. But during the Enlightenment proper, almost nothing was known about biology. As a result, in the eighteenth century, biological science had little real impact on human health and well-being. Even the most modern of Enlightenment biologies were based on primitive assumptions that offered little reliable information about body or mind. But in the nineteenth century, biology grew to maturity. Embryology, the cell theory, the laws of evolution by natural selection, the germ theory of disease, the neuron doctrine, immunology, and transmission genetics—all became modern disciplines. By 1890, with so much new knowledge, the Enlightenment promise of a natural and rational human future seemed real. Human life could be improved, suffering could be stopped, stomachs filled, diseases (sometimes) cured, and minds and morality enhanced.

But the idea of improvement entailed value judgments, and even politically progressive values were at the time linked to their cultural negative: the threat of degeneration. Degeneration was retrogression to a more primitive mode.

Mental illness and/or deviant behavior were signs of mental deterioration that was based in the degeneration of the brain and possibly in reverse evolution. Recast in 1857 in theological-medical terms by the French physician Benedict-Augustin Morel, the concept of degeneration had a deeply religious and moral cast. But from the mid-nineteenth century through World War I, it grew less as a religious and more as a scientific notion. By the 1890s, the popularity of evolution, knowledge of the deleterious effects of inbreeding in animals, and the persistent stress on hereditary causes strengthened the secularization of degeneration.[3]

Degeneration could counter progress with implications of decay, deterioration, and death, and increasingly, the image of a modern rational society that stressed progress, change, and improvement also connoted the threat of backsliding or deterioration. Chamberlin and Gilman describe the concept of degeneration as "an element of a dialectic of thought which . . . embodied something of the structure of an evolutionary reality in which everything moved not only toward a more advanced state but also toward death."[4] The threat was not just moral; rather, the biologizing of degeneration placed it within the very processes of life. Complete integration came with the great German pathologist Rudolph Virchow. Virchow pioneered the connection between cells and organic disease. He argued that the seeds of disease and deterioration emanated from within the very cells that sustain life.[5] Positioned in the cell and linked to pathology, degeneration became an ever-present source of organic decay, a source of hidden weakness that was inherent within the foundation of life and lived alongside productive processes like growth, nourishment, and change. As such, the concept of progress came to entail its opposite—degeneration, a principle lurking within life itself.

The degenerationist paradigm violates the popular myth that medicine began to displace religion and morality in the nineteenth century. Instead, when degenerationism peaked in the late nineteenth century, medicine, biology, and Euro-American middle-class morality formed an alliance that united science and morality at the core of "the normal." The pathological, by contrast, was imbued with immorality. The ideas that enforced what we today often regard as outrageous moral judgments were based neither in poor science nor in pseudoscience. Rather, they were directly linked to modern scientific advances. Degenerationism had entered the fabric of Western science, tightly linked to goals that stressed the progress made possible by science. In some societies, notably German-speaking countries, which led the way in many areas of biological science after 1850, ideologies anchored in medicine and supported by the authority of natural science were so politically powerful that they "provided an alternative to party politics."[6]

Mind, Sex, and Morality in Psychiatry

The years between 1850 and 1900 also saw huge growth in the Western understanding of the brain and nervous system. By 1900, the cell theory had been applied to the nervous system in the form of the neuron doctrine. Staining techniques that permitted the observation of the death and deterioration of parts of the nervous system showed that destruction of lower-level cells (those nearer the spinal cord or peripheral organs) could produce the destruction of higher-level cells (those nearer the cerebral cortex). Somehow neural deterioration worked upward in ways that suggested that "lower" processes could cripple "higher" and presumably more rational processes. There was little understanding of how this occurred in cells, so scientists often looked to clinical evidence from neurological problems associated with "moral" lapses. Prominent among them were the over-consumption of alcohol and unrestrained sexuality. Increasing knowledge of the neurological correlates of diseases such as tertiary syphilis and Korsakov's syndrome (in which the cells of the brain deteriorate and die) seemed to confirm that higher brain centers linked to lapses in will and morality were in jeopardy and at risk of deteriorating.

Early psychiatry viewed mental illness as a symptom of neurological disease, and the idea that morality was at risk was strengthened by the equation of mental illness with brain pathology. Mental illness signaled the loss of reason caused by neurological degeneration.[7] As the symptoms of mental illness progressed, psychiatrists assumed that a patient's brain was dying or at least developing some dangerously large holes, often manifest as moral decline. In this atmosphere, neural degeneration became the prototype for the degeneration of human rationality. So, with little evidence, alongside syphilis, other "diseases" such as epilepsy, masturbation, and the modern American scourge, "neurasthenia," were seen as caused by or producing a degenerating brain. By the 1890s, Europe's most important public health priority was the need to control the potential for human neural and moral degeneracy, as human rationality itself seemed to be slipping away.[8]

Arguably, no one integrated the threat of biological degeneration into psychiatry more effectively than Emil Kraepelin (1856–1926). In 1883, Kraepelin developed the classification system for mental disease that still defines the central divisions of modern psychoses, and he illustrates degenerationist psychiatry's strong emphasis on the negative impact of the environment. He bemoaned the dramatic increase in mental diseases that had occurred in Europe between 1870 and 1900. Coupled with the huge growth of urban populations, the rise of industrialization, and the decline of an agrarian lifestyle, civilization was damaging the psyche through its degenerative effects on brain. Large cities, dirty industrial settings, crowded environments, not enough fresh air, and the unnatural activities that urban people engaged in brought neural insults that led to increases in alcoholism, syphilis, suicide, drug addiction, traumatic neurosis, and neurasthenia. All were, for Kraepelin, manifestations of the degenerative effects of urbanization and industrialization. Degeneration had accumulated with the very progress on which society was built. So, to his nosology of mental pathologies Kraepelin added a new category: "degeneration insanity" (*Entartungsirresein*).[9]

Insanity could be caused directly, when "poisons" such as alcohol and prostitution destroyed the mind of an individual. But Kraepelin believed that the biologically toxic and demoralizing habits of urban life went beyond individuals to harm heredity when degeneration in one person altered the heredity of future generations, thereby threatening the race. Kraepelin, therefore, strengthened a tradition that had linked moral degeneration to the progressive and sustained deterioration of the germ-plasm (the hereditary material carried in eggs and sperm). This meant that in the future, the "number of mental and moral defectives, epileptics, psychopaths, criminals, prostitutes and vagrants" produced by alcoholic and syphilitic parents would be unforeseeably large. Worse, it was aggravated by social welfare reforms that worked against the "natural self-purification of our people."[10] Even middle-class white Europeans, seen then as the superior "race," were in trouble. The hereditary transmission of mental and neural degeneration in this most "civilized" group meant the ultimate decline of humanity as a whole. The paradoxical price of civilization was that in unleashing even more biological degeneration, urban progress was destroying itself; modern life was biologically dangerous.[11]

Sex had long been a part of the mix of morality, degeneration, and disease. Sander Gilman writes: "No realm of human experience is as closely tied to the concept of degeneration as that of sexuality."[12] The threat became even more profound after mid-century, when the new theory of evolution made sex and reproduction central to evolutionary progress. The focus on syphilis and the problem of prostitution, and fears of the deranging effects of masturbation strengthened the connection. They aggravated an ongoing process of devolution. By 1890, cretinism, mental retardation, syphilis, epilepsy, prostitution, criminality, and Krafft-Ebing's various "perversities" were all seen as forms of biological degeneration accompanied by irrational or depraved sexuality.[13] These encroachments on middle-class standards reinforced existing racist, classist, and sexist views of humanity. But now they were empowered by the authority of biological psychiatry. White middle-class manhood, the ideal of civilized progress, will, and reason, was threatened. Intermixing with "inferior races," "seduction" by lower-class prostitutes, "diseased" expressions of sexuality seen in masculine women, Jews, Africans, avant-garde artists, and the "feebleminded" had to be tamed and eliminated. Again, we can turn to Krafft-Ebing; along with the psychopaths and the perverts, included in his list of those suffering from neurologically based mental degeneration were "eccentrics, misanthropes, political scatterbrains, ardent social reformers, religious fanatics and . . . otherworldly scholars and artists."[14]

Developmental Approaches to Sexuality: Reproductive Endocrinology

These views dominated the late nineteenth century during the first great age of biological psychiatry, before the rise of hormones. But after 1900, a new type of cause transformed the biologizing of sexuality and conceptions of the human constitution—the hypothetical inner secretions, which in 1904 were called "hormones." Hormones were controversial substances even before they were given their modern name. They smacked of the ancient humoral theory of the Greeks, which had finally been overthrown by the recognition that the brain and nervous system coordinated mind and body. By 1900, no one doubted the power of the nervous system. But clinical evidence also suggested pervasive chemical influences over mind and body that could affect human character in ways that did not depend on known brain processes. The question of neural versus chemical control was the central issue in the pioneering years of endocrinology between 1870 and 1910, and nowhere was the problem more pressing than with the biological understanding of the chemicals influencing the sex drive. Strong evidence came from deficient or overactive glands, organs like the adrenals and the thyroid. Along with several disease susceptibilities, thyroid deficiency produced a large skull, short plump hands, small expressionless eyes, inactivity, lethargy, and numerous sexual disorders (undeveloped or underdeveloped genitals, dysmenorrhea, and the absence of a sex drive).[15] The range of seemingly unrelated effects raised interest in what kind of process could alter so many dispersed organs and systems. And why was sexual dysfunction produced by problems with glands, like the thyroid or the adrenals, whose secretions seemed unrelated to sexuality? Feeblemindedness was often involved, raising the question of how brain science would have to be modified to accommodate influences with such odd complexes of dispersed glandular symptoms.

By 1910, endocrine hormones had become the likely candidates. Glandular deficiencies and excesses could influence a wide range of clinical symptoms involving appearance, disease susceptibility, temperament, and personality. Research on the thyroid gland showed that hormone secretions could alter the body separately from the nervous system. But physiologists who privileged the nervous system maintained that secretions could not act without direct involvement by neurons, especially when mental symptoms (like feeblemindedness or an overactive sex drive) were involved.[16] Part of the confusion stemmed from the odd anatomy, or actually the anti-anatomy, of the endocrine glands. Endocrine glands are ductless glands. They carry their secretions not in systems of ducts or tubes through which the chemicals flow, but by depositing their secretions directly into the circulating blood. For this reason, their products, the hormones, travel everywhere that the blood flows, leaving no anatomical traces of their influence. Because of their ductless character, the chemicals and their actions could not be traced as other secretions could, using anatomical connections between the glands that produced them and the organs that reacted to them. Plus, because the secretions could reach any tissue that the blood did, their influence could be pervasive, affecting many systems simultaneously. These were both strengths and weaknesses. They made it possible to account for the diffuse effects of endocrine hormones; but in the days before the synthesis of the chemicals, those diffuse effects made it difficult to say that they were there at all, much less to indicate where their influence originated.[17]

As a result, demonstration of the action of the hypothetical chemicals depended more on outcomes than on known chemical processes.[18] And although pathological manifestations were often powerful, clinical data made it extremely difficult to separate one kind of cellular influence from another. The responsible chemical could come from anywhere, and its effect had to be inferred from a wide range of functional outcomes. Animal experiments were required to separate the effects of chemical secretions from the possible influence of nervous pathways and to infer the glandular source of the chemicals. These began consistently around the turn of the century, when glandular transplant techniques were pioneered with the thyroid gland. Because the transplants severed connections to any proposed neural source, transplant studies became the dominant methodology supporting the power of internal secretions. Between 1900 and 1910, a confluence of clinical, pathological, and zoological studies strengthened the idea that inner secretions could determine bodily and mental characteristics independently of the nervous system.[19]

The "Vienna School" of Reproductive Endocrinology

By 1910, many scientists accepted the idea that hormones transmitted via blood could alter brain and body without mediation by the nervous system. But an understanding of the hormonal basis of sexuality lagged behind, perhaps because neural interpretations persisted in the face of a powerful psychic correlate: the sex drive.[20] It remained to extend the concept of autonomously acting hormones to mammalian sexuality through successful gonadal transplants on mammals. These experiments finally occurred in a milieu that rejected the tradition of degenerationist approaches to sexual diversity and stressed instead regeneration and development; they were done in the Vienna laboratory of Eugen Steinach.

Steinach was the son of a middle-class physician. He had moved to Vienna from his hometown in the Austrian Alps to study medicine at the university. In

1885, he left Vienna to build a career, first in Innsbruck and then for many years in Prague. By 1910, the year Tandler became head of the university's Anatomical Institute, Steinach was negotiating a return to Vienna. He came back in 1912 to establish his laboratory and head the Department of Physiology at the Institute for Experimental Biology (*die biologische Versuchsanstalt*). The Institute was an independent laboratory, founded and owned by zoologist Hans Przibram. With two friends, Przibram had purchased an old aquarium built for the 1872 World's Fair in the Prater, Vienna's huge public park. Przibram rebuilt the space into a state-of-the-art biological research facility that opened in 1902.

The idea of biological malleability, plasticity especially apparent in the development of organisms, was central to the Institute's approach. It supported a worldview more in accord with the developmental fluidity of flexible hormones than the rigid "solidity" that dominated conceptions of nerves.[21] That view had already been advocated by a giant of Austrian biology, Sigmund Exner von Erwarten. Born in Vienna to an extremely influential and scientifically oriented family, Exner was a neurophysiologist who had studied with the renowned German physiologist Ernst Brücke. Along with Hermann von Helmholtz, Brücke had ushered physiology into its modern phase, and to ensure his future, Brücke sent the young Exner to Heidelberg to work with Helmholtz. On his return to Vienna, Exner rose quickly. On Brücke's death, Exner was named his successor as director of the University of Vienna's Physiological Institute. By 1900, he was arguably the premier biologist in Vienna.[22]

Exner rose at a time when three of the great coordinating systems of the body, the neural, the genetic, and the hormonal, were being defined. The term "gene" was coined in 1909, only five years after the term "hormone." But how these systems influenced one another, and whether each acted autonomously were questions still hotly debated in the domain of sexuality. Did hereditary particles determine whether one was male or female, or were the glands autonomous in determining an individual's sexual attributes? If hereditary particles were decisive, when would they act? And could hormone secretions counter their impact? Exner framed the Austrian response to these questions, when, just as Przibram opened his Institute, he turned his attention to the sex drive.[23]

Exner believed that most characteristics of individuals are hereditary: they result from instructions that are predetermined in germ cells. But the secondary sexual characteristics were different. The sex drive and secondary sexual characteristics, including those in brain, were not produced by genes. They were not predetermined; rather, they developed flexibly under the influence of the sex glands. "There is therefore no organization in the germ cells which determines that a child of a given age has a beard. Rather, there is an organization that determines that in some way normally functioning testes exert an effect on the facial hair."[24] Citing Steinach's 1894 study on rats, which supported the autonomous action of the chemicals. Exner argued that the germ-plasm had relinquished direct control of sexuality to the secretory products of the gonads. Sexuality was "plastic." The sexual secretions even seemed to alter the chemistry of brain, conditioning its neurons. When they reached the central nervous system, they transformed psychic states that included one's emotional tone (*Gefühlstimmung*), the sensations serving desire, and the impulses of attraction and affection. The effects were so powerful, he said, that one could view "the sex drive and love as a kind of intoxication."[25]

The Methodology of Glandular Transplants

Perhaps because he was primarily a neurophysiologist, Exner's rejection of a strictly neural and rigid account of sexuality was significant in shaping early Central European endocrinology. Between 1905 and 1930, an Austrian/Czech school of endocrinology led by Tandler, Steinach, and Artur Biedl grew.[26] Viennese and Prague endocrinologists stressed the independence of hormonal control from and its interrelation to brain through the pituitary gland. Developing and extending that approach required the regular use of transplants that severed a gland's connections to neural tissue, but left the organ healthy and able to secrete its hormones. Testicular transplants had been tried before, but they rarely worked, and when they did they usually showed only the impact of the secretions at autopsy, after an animal died. This left meager, if any, basis for inferences about the sex drive.[27] Experiments on the ovaries fared somewhat better. Viennese gynecologists Emil Knauer and Joseph Halban began ovarian transplants on rabbits and guinea pigs in the 1890s. When transplants were performed on young females, the animals' internal reproductive organs were normal, and they showed normal uterine maturation. Nerve tracts did not appear to control the ovaries, making it likely that the glands were the source of significant secretions. But Halban, whose work was most influential, denied the causative power of the secretions and argued that they were merely protective in nature.[28]

Addressing the development of the sex drive required a new kind of transplant experiment. Because so many processes were involved, short-term transplants were unlikely to reveal all the developmental outcomes. And timing in development was important. In males, for example, pre-pubertal castrations had different effects than post-pubertal castrations. The most useful transplants had to be extended in time to enable scientists to assess sexual functions across months or years, as the beasts lived and changed, and not merely in fixed dead tissues. Such chronic experiments would show how the chemicals altered development to influence behavior and drives later in life, at puberty and even in adulthood. They might even show how these influences changed in response to the presence of different glands producing different secretions, proving, as

Exner suggested, that they were flexible. It was Steinach who perfected such chronic testicular transplant experiments in mammals. He was the first to demonstrate the developmental flexibility of sexual characteristics that could be permanently altered by different gonadal hormones acting on very young mammals. He stressed the "psychic" significance of the secretions, examining changes in drives and behavior that persisted almost throughout an animal's entire lifetime.

Steinach and the Double Gonad

For many years the biological study of differences between the sexes had been dictated by the characteristics of their gametes—the sex cells produced by males and females. Mammalian and avian eggs were large and unreactive, they appeared "passive," and they contained the nutrients required to nurture the developing young. Sperm, on the other hand, were active, quick, and competitive; they raced for position, and they played no obvious role in nurturing. It was easy to assume that they might even be aggressive. These were just descriptions of the cells produced by meiosis prior to fertilization. But for more than half a century, the implications of gamete structure were taken much further. In a massive cultural projection from cells to individuals, largely male scientists assumed that the characteristics of individuals' gametes also dictated the social demeanor and the psychic sexuality of adult men and women.

Like their eggs, women were assumed to be passive, nurturing, and unreactive; men, like their sperm, were considered active, competitive, and aggressive, and they were believed to take little interest in their offspring. Probably because true action implied volition and the exercise of will, the dimensions of active and passive were also fused culturally with the presumed rational and decisive nature of men as compared to the emotional and nurturant dispositions said to characterize women. The sexual attributes of men and women became projections of their gametes, dictating even mind and personality. The liberal pathologist Virchow expressed the views of over half a century when he wrote, "The female is female because of her glands of generation. All her characteristics of body and mind . . . in short all that we admire and honor as feminine about the true female is dependent on the egg (*Eierstock*)."[29] Men controlled and were controlled by their brains. But a woman's being, body and mind, were given in the character of her eggs. Importantly, the future of the species itself required that women's sexualities be dictated by the reproductive functions of their eggs. In his work *La donna delinquente, la prostituta e la donna normale* (Criminal woman, the prostitute, and the normal woman), Cesare Lombroso wrote this of love in "normal" women: "the love of a woman is at its base nothing other than a secondary quality of motherhood. All the feelings of affection that bind a woman to a man emerge not from sexual impulses, but through acquired instincts from submission and sacrifice."[30] Passive gametes and motherhood, not assertive sex drives, defined women.

By about 1906, Steinach and Tandler began to frame sexuality in a radically new light. They independently extended a theory that rejected the tradition of femininity based in eggs, childbearing, and motherhood. Their approach instead separated reproduction, as a means to generate offspring, from sexuality, as an independent expression of anatomy and psyche, which they extended to women as well as to men. Both men championed the idea, first published by the French anatomists Pol Bouin and Paul Ancel, that there were really two glands in the gonad: one functioned in reproduction proper; but it coexisted with a micro-anatomically distinct gland that functioned to control sexuality. Because this hypothesis placed the power to shape sexual attributes, not in the gametes but in the separate cell group making up the second gland, the French view challenged the assumption that eggs dictated the psychological attributes of females. Like men, women had a sex drive, and like men, it was based in the psycho-physical reactions of the brain and body to hormones produced by a separate structure in the gonad. Between 1910 and 1920, the application of this hypothesis to mammals rested firmly on the research made possible by Steinach's chronic infant transplants.

The Interstitial Cells

In mammalian testes, sperm are formed in a complex of coiled tubes called the seminiferous tubules that fill the testes in a densely packed net. Since the 1880s, it had been known that three distinct cell types were intermixed in this net. The first group was the Sertoli cells, which resided in the semen-producing tubules in which sperm grew; the second included the generative cells themselves, mature sperm and the cells that gave rise to them at the base of the inner surface of the tubules. The third group was the interstitial cells, large fat-containing cells located in the spaces between the tubules. The prevailing view at the turn of the century was that the interstitial cells were connective tissue that perhaps functioned to provide the nourishment required for spermatogenesis in the other two cell types. Many did not yet accept the power of inner secretions in shaping sexuality, but even those who did usually believed that those secretions were the product of the sperm themselves. This began to change just after the turn of the century, when French anatomists proposed that the interstitial cells were a separate endocrine structure within the testis.

Cellular anatomist Franz Leydig first described the interstitial cells in 1850.[31] In 1903, Ancel and Bouin termed the cells a *glande interstitielle,* and by 1910, they had published several papers suggesting that the sexual secretions that came from the testes were products of the interstitial cells, not the sperm or the

Sertoli cells, as others argued. The double gonad hypothesis effectively separated the cellular mechanisms of reproduction from those controlling sexuality. One portion of the gland, the generative part, produced gametes, the sperm and their progenitors. They functioned in reproduction. The inner secretions, which were produced by the regions composed of interstitial cells, controlled sexuality proper. The testes were two distinct glands in one. Working independently, Steinach and Tandler had both examined the testes, and both based their approach to the biology of sexuality on the double gonad hypothesis: hormones secreted by the interstitial cells controlled sexuality, while Sertoli cells controlled gamete production and, therefore, reproduction.

Between 1905 and 1916, more and more evidence confirmed the sex-determining power of secretions produced by the interstitial cells. Tandler completed some of the earliest work supporting the interstitial cell hypothesis. He had done pioneering research on human eunuchs, but he also explored the cellular basis of hormone secretion in the testes of deer and moles.[32] In studies begun in 1907, Tandler and Grosz showed that radiation produced a proliferation of the interstitial cells in male deer, while their sperm atrophied and even died. Despite damage to the sperm, however, the animals' secondary sexual characteristics were unaffected or exaggerated. This could not occur if secretions from the sperm regulated sexuality.[33] Again with Grosz, Tandler also showed that natural seasonal cycles of reproductive activity in the testes of moles were correlated with inverse cyclical changes in the two cell types. Over two years, as males bred, the sperm-producing cells gradually atrophied, while the interstitial cells began to proliferate. Months later, when the sperm-producing portion of the testes redeveloped and sperm grew, the interstitial cells were again greatly reduced. The two cell types fluctuated in inverse relation to one another (see fig. 2.1). The authors concluded that the interstitial cells determined the sexual maturity of individuals and the growth of their germ cells. "In the . . . development of the interstitial cells, we see the determining cause for the next spermatogenesis."[34]

But Tandler was more an anatomist than an experimentalist, and he had almost no expertise with the transplant technique. It was Steinach who produced the most convincing and most controversial research supporting the double gonad. Extending the idea to mammalian sexual development became his central interest, and by 1926, his work offered the largest body of evidence supporting it. Critics referred to the "Vienna school" of reproductive endocrinology to indicate that camp of reproductive endocrinologists, by then identified with Steinach, who promoted the separation of reproductive and sexual functions in the gonad and who built a notion of "hormonal bisexuality" that linked maleness and femaleness to chemicals produced by the interstitial cells, not the gametes.

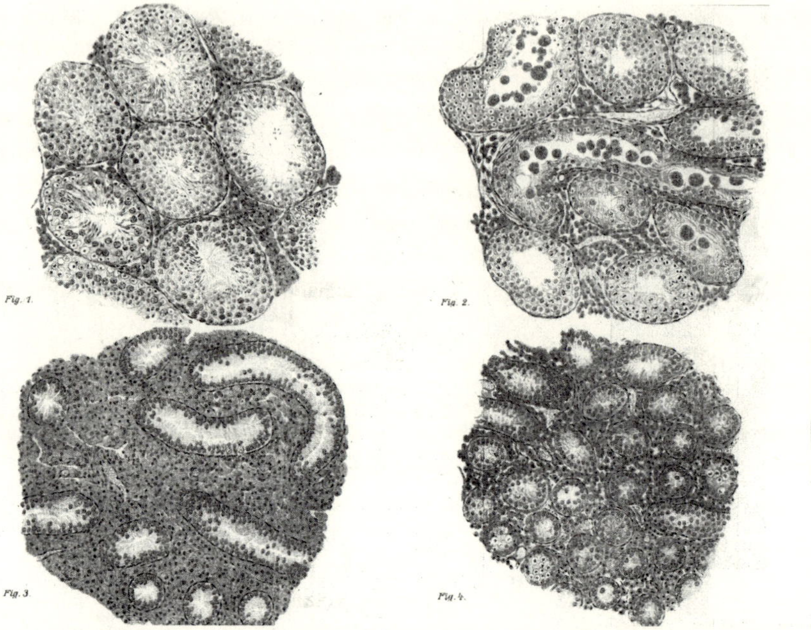

FIGURE 2.1. Cross sections of mole testes showing the interstitial cells and sperm at four different points in the seasonal cycle of reproduction. The interstitial cells are interspersed among variously shaped tubules in which sperm are produced. "Fig. 1" (April, at peak breeding) sperm are being produced near the centers of the tubules with few interstitial cells evenly spaced around the tubes; "Fig. 2" (end of May, after breeding) only sporadic sperm, usually without heads, and more interstitial cells; "Fig. 3" (end of July) depicts many more interstitial cells, now with clear cell bodies, and no sperm in the irregularly shaped tubes; "Fig. 4" (December) the tubules are more regularly shaped and again evenly surrounded, but by fewer interstitial cells, as sperm begin to develop. Over a year the two cell types change in inverse relation to one another.

Table XVI from Julius Tandler and Siegfried Grosz, "Über den Saisondimorphismus des Maulwurfhodens," *Archiv für Entwicklungsmechanik* (*Development Genes and Evolution*) 33 (1911–1912): 297–302. Reproduced with the permission of Springer Scientific.

Steinach's Chronic Transplants

Steinach perfected same-sex and cross-sex chronic transplants of the gonads in rats and guinea pigs. The experiments differed in two ways from most research that had come before. First, framing his question in developmental terms, Steinach began with very young animals, pioneering the combination of castration plus transplant in infants in which no obvious sexual differentiation had yet occurred. Rats develop sexual differences in anatomy and brain not prenatally, but around the time of and even after birth. Because many body and brain structures are not developed at birth, experiments begun this early could show the malleability of sexual development in ways not possible in older animals. Second, his experiments were chronic, that is, they extended to well after the time of puberty. Animals were castrated days or weeks after birth, and the same or different gonads were replaced in a new location in their bodies. The animals were then nurtured, reared, and measured for as long as two years to reveal the unfolding of differences produced by the replaced glands that could be traced well into adulthood.

In Steinach's first transplant study, infant male rats underwent transplants in which the gonads were removed and replaced into the same animal's abdomen. Forty-four survived, healed, and grew to adulthood. Animals with transplanted testicles showed normal sexual anatomy and normal sexual behavior. Anatomically, they were fully developed males, and they behaved as normal males. Because the repositioning of gonads severed all connecting nerves, the study helped establish the impact of blood-borne chemicals on the development of masculine sexual characteristics in mammals. The occurrence of normal adult sexual behavior indicated that secretions from the replaced testes, not nerves, controlled the gland's effects on sexual interest and behavior. The experiments strengthened the rejection of exclusively neural interpretations of sexuality.

Steinach also gathered cellular evidence to support the double gonad hypothesis. Microscopic analysis of cells in the transplanted testes showed no evidence of living sperm growing in the animals' sperm ducts. The interstitial cells, on the other hand, were either normal or greatly increased in number. Steinach cited Tandler's demonstrations that secondary sexual characteristics and behavior depended, not on sperm tissue but on the interstitial cells. Tandler had apparently even helped with the work: Steinach thanked him for performing some of the autopsies and microscopic examinations. Steinach had, with a more definitive methodology, reproduced Tandler's result and extended its significance to sexual development. "Not simply the maintenance, but the entire development of masculinity and the growth of the secondary sexual characteristics were dominated by the activity of the inner secretory tissue."[35]

In these transplants, the organs were removed and replaced in a different location in the same individual. Steinach also developed transplants in which he replaced the glands in different individuals of the same species. They allowed him to ask if the impact of the hormones was sexually specific. If so, they carried a sexual "principle" that potentially could transform the attributes of males into those of females, no matter which sex held them. In a series of works, which appeared between 1912 and 1916, Steinach castrated young male rats and guinea pigs that had shown no signs of sexual development and gave them ovaries. With this method, he could determine whether female glands in males led to female or to male sexual development and behavior. If the first, the effects of the glands must be sexually specific: the glands and their hormones, not the animals or their heredity, were male or female. If, however, male features developed in males in the presence of healthy, secreting female glands, those glands must be overpowered or neutralized by some other principle of maleness possibly present in the germ-plasm. This would indicate that Exner had overstated the case; something less developmentally flexible than hormones shaped sexual attributes. Further, the cross-sex transplants turned on the plausibility of the separation between the generative and secretory parts in the double gonad. If gametes controlled both the nature of sexuality and the character of the secretions, the question of sexual specificity was settled. The secretions and their effects would match the character of the gametes producing them, and effective cross-sex transplants would depend on the presence of healthy gametes. But if sexual characteristics were the result of hormones produced independently of the gametes, and if these hormones were sexually specific and acted flexibly, functioning ovaries should feminize developing males.

And feminize they did. Anatomical results showed a near complete transformation in secondary sexual characteristics dictated by the nature of the transplanted gonad.[36] Males' external sexual anatomies (for example their very small penises), hair growth, accessory sexual structures, body size, and skeletal structure were feminized by the implanted ovaries. Steinach emphasized the effects of the implanted gonads on his most precise measure: changes in skeletal structure revealed by x-ray radiation. The skeletons of adult males implanted as infants were considerably smaller than their normal and just-castrated brothers and very similar to their sisters. Males with ovaries even developed functioning nipples and milk glands that were virtually the same as those of females. The cellular structure of the transplanted ovaries further suggested that the follicles were functioning normally in males. Normal ova were produced (though they did not reach their typical size in females), and some males formed corpora lutea (the source of progesterone in the burst ova). In males with the most successful implants, the ovaries looked so good that Steinach went so far as to suggest the possibility that males could sustain pregnancy!

The parallel study on females appeared in 1913.[37] Females' ovaries were removed in infancy and they were given their brothers' testes. They developed very masculine characteristics at puberty. They were larger than their sisters, their body structures were male, some had completely closed vaginal openings, and they had male-like fur. Though the cellular structure of the gonad was less clear in females, Steinach generalized from the males: it was the interstitial component of the gonad that was responsible for both the masculinizing patterns in females and the feminizing patterns in males given cross-sex transplants.

Steinach concluded that the interstitial cell's secretions were sexually specific and flexibly linked to producing masculinity and femininity. Female gonads formed female characteristics in males, and male gonads formed male characteristics in females. In addition, however, in both sexes, masculinized females or feminized males had some features that were inhibited by the opposite gonad (for example, growth in feminized males) and some that were exaggerated (for example, nipples in feminized males). So, the glands normally exerted two activities; normal secretions of, for example, female gonads in males could inhibit as well as excite a feminine outcome.[38] Most important, sexuality was not given in the gametes (or their germ-plasm). It was flexible. Masculinity or femininity could appear in either males or females. Steinach concluded: "Neither the secondary somatic nor the secondary psychic sexual characteristics are unchangeable or predetermined from the start (*ab ovo*). They can be transformed, that is to say, changed."[39] The sexual transformations were so complete that they must develop from a sexually indifferent plan: "The tendency (*Anlage*) of the embryo is neither single-sexed nor dual-sexed, rather it is asexual or indifferent."[40]

Throughout, Steinach was especially interested in the "psychic" findings. Even his early work addressed not just the bodily manifestations of sexuality but neural ones, as well. In 1894, he studied the impact of secretions on the nervous systems of frogs, animals in which the neurophysiology of the copulatory reflex was known. Castration eliminated the power of a stimulus to release a natural copulatory reflex. But within forty-eight hours of a hormone injection, stimulation worked again; the reflex was restored to its maximal potential. Hormones had altered the nerves and the sex drive. Though he had no neural data on mammals, combining his inferences from the rats' behavior with his findings on frogs, Steinach interpreted the "psychic" effects of the mammalian hormones in terms of their impact on the brain. He proposed that the first developmental impact of the secretions occurred in the central nervous system. He termed the process the "'erotization' (*Erotisierung*) of the central nervous system," noting that it could long outlast a decline in its original cause, the circulating hormone.[41]

The behavior of his transplanted animals consistently supported these inferences, and that support probably aggravated scientific debates about the cellular structure of the double gonad. Ovary-implanted males showed no signs of a male sex drive. The hormones from the female gland, he concluded, had produced an "erotization" of the central nervous system in a female direction. This was also apparent in the adult behavioral interactions between feminized males and control males. The implants triggered strong sexual pursuit of feminized males by the untreated males; the males with functioning ovaries were being treated like females. And ovary-implanted males were sometimes sexually receptive, reacting to untreated males with lordosis (*Schwanzreflex*), the characteristic female rat receptivity posture. To fend off the advances of untreated males, rather than fighting, as male castrates might, males with ovaries often showed a female defensive reflex. Describing the psychic effects in males with ovaries, Steinach wrote: "The operated animals acquired a female being, a female mood," one whose natural authenticity was indicated by the strong sexual reaction they awakened in untreated males.[42] Masculinized females showed parallel male sex drives—their "psychosexual" transformation, said Steinach, was even more extreme than their anatomical changes. They could distinguish receptive from nonreceptive females; they mounted and pursued receptive females; and they fought with males. Citing Exner's analysis of the relation between hormones and the nervous system, Steinach suggested that in higher life forms, erotization by the interstitial cells was the likely basis of individual differences in psychic and somatic sexual dispositions. Years later, in his scientific autobiography, Steinach overextended the significance of neural erotization to the person, stating that the "mutually interdependent" endocrine system is responsible for "psychic makeup, emotions, instinct, and desires—in a word, our personalities."[43]

Bisexuality

In 1916, Steinach published the even more striking results of dual transplants. These were experiments in which male and female glands were both implanted into castrated males or females days after birth. Previous cross-sex transplants without prior castration had not worked, despite the evidence for hormonal specificity. Steinach interpreted this as evidence of hormonal "antagonism": somehow the intact original gonad had gained the developmental upper hand and prevented the transplanted gonad from exerting its effect.[44] What would happen if both glands were simultaneously introduced into infants that had been "neutralized" (castrated) at a young age? Days after birth and before the natural gland had exerted much influence, Steinach transplanted ovaries and testes into the same castrated males. A small percentage of the animals became hermaphrodites; he called them *Zwitter* (double-sexed).[45] Though this was not the most common outcome (only 20 percent became double-sexed; others were just feminized), the Zwitter showed developed characteristics of both sexes. Compared to their untreated and castrated brothers and to their untreated and castrated sisters, they were male in size, fur density, and genital structure. The inhibitory influence of the female glands was ineffective in the presence of the male glands when both could act from a young age. But, while the Zwitter had well-developed erectile penises, their penises developed alongside milk-producing teats. Psychically too, the animals were mixed. They fought with strange animals and showed male sex drives; but they were erratic. Sometimes they were afraid and ran from a strange male; sometimes they behaved like receptive females. The feminine and masculine "psychic" characteristics did not coincide; rather they seemed to alternate with one another, both recurring periodically. Finally, when the grafted female gland was removed in males after both male and female characteristics were manifest, the psychic and bodily female attributes disappeared and the manifestations of female sexual behavior never returned. The reverse happened in females when grafted male glands were removed. Steinach expressed surprise that the central nervous system was so reactive to periodic fluctuations in the hormones produced. He speculated that there must be a central substance that could be transformed by gonadal hormones.

Cellular data on gonad structure confirmed hermaphroditism in the gland itself. When both transplants were grafted successfully to the same muscle, an "Ovotestes interstitialis" resulted. This was a gland in which male and female interstitial cells were interspersed with one another in neighboring "islands."[46] Microscopic cell analysis confirmed what transplant and radiation had suggested: the changes in the bodily and behavioral signs of sexuality were accompanied by huge growth in the "incretory" part of the gonad, the interstitial cells, which Steinach by now termed the "puberty gland."[47] Animal studies inducing the early onset of sexual maturity confirmed that this was an accurate label. "The result was the same in each [gland] explored: normal development of spermatogenesis, normal expansion and opening of the sperm canal, but exaggerated (*übernormale*) development of the puberty gland, which reveals itself . . . in the entire [testis] through a quite comparable thickening and extension of the interstitial tissue complex."[48]

Human and animal hermaphrodites had been of interest to biologists for decades, and debates about them and about the causes of sex determination were complicated by the discovery of sex chromosomes in 1905, and the description of sex-linked hereditary transmission in 1910. Though Steinach said almost nothing about genes, the growing emphasis on sex chromosomes intensified debates about the relative roles of genes and hormones in determining sex.[49] In the same year that Steinach produced his Zwitter, Richard Goldschmidt also stressed development in research on the sexually intermediate forms produced by different hybrid crosses of gypsy moths. Certain crosses produced females with various degrees of mixed male and female attributes. Goldschmidt termed them intersexuals, and though he stressed the importance of chromosomes, he also challenged the idea that they dictated two discrete classes of attributes (male versus female). Instead, Goldschmidt, like Steinach, stressed physiology and left room for numerous intermediary forms determined by the quantity of the sex-determining substances.[50] Indeed, Steinach's view of the hormonal antagonism produced by transplanting both mammalian glands into the same individuals also emphasized developmental timing. Development was the key, and by 1920, Steinach's research had consistently confirmed a distinct gland in the mammalian gonad that was composed of interstitial cells and that controlled many manifestations of sexuality. Different cellular processes governed sexuality and reproduction, and those shaping sexuality showed enormous flexibility.

The implications for traditional conceptualizations of masculinity and femininity were profound. If the gonad was truly a double gland, sexuality was separate from reproduction and a woman's reproductive specialization did not determine her femininity—not if the interstitial cells played similar roles in the ovaries and the testes. Moreover, if the interstitial cells played parallel roles in males and females, they implied, as Biedl had suggested in 1910, a natural cellular condition of "physiological hermaphroditism" in mammals.[51] The concept implied that the potential to be male or female existed in the hormone-producing cells, and, given the developmental flexibility seen in Steinach's transplants, that bisexual potential could be shaped in many directions. So not only was sexuality separate from reproduction, but early in development the physiology of masculinity and that of femininity might be equally represented in males and females.

Almost as if to further challenge the dominant cultural categories, Steinach pursued the notion of bisexuality in the late twenties. With his student H. Kun, he turned to the pituitary gland (the "master gland" that in mammals hangs off the base of the brain) and injected animals with hormone extracts to show the bisexual effects on males and females of the same naturally occurring hormone. Like others, Steinach had known for years that the pituitary gland was also somehow involved in sexuality. When Steinach and Kun injected infant female guinea pigs with extracts of pituitary hormone after their ovaries were removed, the young females were still "female," but they also developed large penises, male vocalizations, and masculine behavior. No masculinization occurred in females with their ovaries intact. But without them, pituitary injections triggered the development of masculine *and* feminine characteristics; natural hormones could sustain "experimental hermaphroditism."[52]

Surprisingly, injecting corpus luteum extract into castrated female pups also masculinized them. The authors concluded there was a natural "masculine" hormone produced in the female corpus luteum, and to confirm it they tested the hormone on males. Castrated control male rats showed poorly developed penises, while castrated males injected with the corpus luteum extract from females developed penises that were indistinguishable from those of untreated males. Steinach and Kun concluded, "The luteal tissue also produced an incretion whose effect corresponds to that of the testis hormone." It could masculinize either females or males. The only explanation seemed to be that the ovary naturally produced two hormones: one that shaped female sexual characteristics, and one that shaped male characteristics in either sex. "There exists in the ovary a condition that therefore can be called *hormonal bisexuality*."[53]

The transplant studies had shown that mammalian sexual anatomy, sexual attributes, sexual behavior, and even sexual preference were not the result of neurons or sex chromosomes. They were fluid, literally, and very malleable. By manipulating hormones, Steinach had confirmed Exner's assertion, shaping persistent and pervasive developmental changes in masculinity, femininity, and their combination. Sexual characteristics were not fixed at conception. Hormones produced them, and they did so independently of neural control in a flexible process of development in which the "indifferent" Anlage of the embryo was formed by secretions. Many other endocrinologists agreed. Commenting on the significance of Steinach's transplant studies, Copenhagen endocrinologist Knud Sand said that they had "placed all others in the shadows."[54] Summarizing the significance of the double gonad theory, Steinach later wrote, "Perfect specimens of one single sex are in reality theoretical ideals; a complete man is as non-existent as a complete woman. . . . It is possible to detect in every man some, though possibly minute, traces of femininity, and in every woman some slight attributes of masculinity. . . . Between a real man and a real woman there are innumerable others."[55]

Steinach's results were striking. Males suckled young and became receptive to other males; females developed penises and chased females. Natural female hormones could masculinize females or males. His conclusions had profound implications in a culture dominated by assumptions of rigid differences distinguishing men from women. Vienna's superficial sense of pleasure and prosperity was notoriously accompanied by great misgivings about self and identity, especially as related to sexuality.[56] Steinach's research contributed significantly to a growing list of threats that aggravated this dis-ease. The boundaries had already been blurred by changes in culture and society, but this threat was coming from within science, once a safe modern bastion of rigid sex differences. Steinach's research had shown that reproductively, a woman could still be a woman. But, sexually, she not only had a sex drive, she might show many of the masculine attributes that could well be absent in a man.

Steinach and Sexual Reform

Steinach's work challenged the boundary between men and women (though not necessarily that between masculinity and femininity) and shattered the biological foundation for presumed sex differences based in gametes. Not surprisingly, the international community of sexual reformers took notice. In 1930, the World League for Sexual Reform, founded in 1928 and led by the German homosexual reformer Magnus Hirschfeld, met in Vienna. The group had been active for almost a decade, and the meeting was a who's who of sexual reformers. The International Board included Biedl, Harry Benjamin, Rudolf Goldscheid,

Rosa Mayreder, Julian Marcuse, and Helene Stöcker; the honorary presidents were Havelock Ellis and August Forel; Hirschfeld presided at the meeting. As the scholars were welcomed to Vienna, homage was paid to the revolutionary impact of the Viennese pioneers, Freud and Krafft-Ebing; but the proceedings show that Steinach and Tandler were now among those laying the foundation for modern sexual reform. The movement had come to greatly respect the significance of their work. Hirschfeld himself had invited Tandler to participate in the League's second congress in 1929. Hirschfeld then wrote: "The international reputation of your name and your work would surely raise considerably the prestige and success of this Congress, which in cultural perspective will be most meaningful."[57] Tandler did not attend; but a year later he spoke at the Vienna Congress. The Vienna program contained a segment on inner secretions; and several papers repeatedly referred to Steinach's work. Josef Novak, then a professor of gynaecology and an expert on female anatomy, wrote that the interstitial cell evidence had "blurred the sharp boundary between masculinity and femininity."[58] Steinach did not attend the meeting; but he invited the participants to tour his laboratory and see his demonstrations of sexual flexibility. The invitation suggests that Steinach, who carefully guarded his scientific reputation, was comfortable aligning himself with the principles behind the League's founding, and they with him.

The League's foundational goals were listed in the published volume of its proceedings. They stressed sexual equality and "the sexual rights of women," birth control, enlightened approaches to "intersexual variations" (translated in the English abstract as "sexually abnormal persons"), and race betterment.[59] Steinach and Tandler's endocrine work was considered biologically foundational to each of these goals. It had shown that diverse expressions of sexuality were not the result of neurologically degenerative disorders. Chemicals that were a normal part of sexual development could dictate natural variations in sexual preference and sexual behavior. From this perspective, the degenerationists were wrong. Sexual diversity was not, as Krafft-Ebing originally thought, perversity, depravity, or madness. Developmental processes in various "perverts," including "psychic hermaphrodites," were not fundamentally different from—indeed were continuous with—those in sexually staid and socially "normal" individuals. They involved a mix of circulating hormones, which could be shaped and altered in a number of ways, and which instilled masculinity and femininity in the same individual. For these reformers, the reproductive endocrinologists of the "Vienna school" had replaced a sexual biology of degeneracy with a flexible biology of development that had little to do with gametes. The nature of the gametes they produce, while critical for making a new individual, did not define individuals' sexuality, even their psychic sexuality. It is unlikely that any other program of research in early twentieth century endocrinology had so effectively challenged prevailing social standards.

3

Paul Kammerer and Flexible Heredity

In the mid-1920s, the American psychiatrist Karl Menninger published a review of a book entitled *The Inheritance of Acquired Characteristics.* The book dealt with heredity, and Menninger, who explored humanity's destructive tendencies, acknowledged that he had insufficient expertise to assess the book's scientific merit. His assessment was based not on the science but on his psychiatric analysis of its author, Paul Kammerer. Menninger framed the review as a dialog between the missing patient (Kammerer), whose "symptoms" were reflected in the pages of the book, and his unsought-after therapist (Menninger). The patient, the psychiatrist concluded, was clinically paranoid. Through his writing, Menninger saw Kammerer's "exaggerated defense against projected hostility" and his "commensurate exaggeration of the ego." Menninger said, "there is a semblance of presenting the opposite side of the question but it is not well done, for this the paranoid individual cannot well do." On these grounds the author's science could not be trusted. His mind was too unstable. Of the ideas, Menninger said: "It is impossible to discount sufficiently the prevailing paranoid, and hence unscientific tone of the thing."[1]

This was the same Kammerer who, in that same year, was judged a pleasant dinner companion by H. S. Jennings. Jennings had taken a less personal look at the science. He liked what he saw, and he suspected that Kammerer's rivals were unjustly judging him. The two assessments of one man's scientific work could not have been more different. They reveal the wide range of reactions that Kammerer's contemporaries brought to his work and his personality. Menninger's hasty judgment and psychiatric hubris, based on no direct knowledge of his "patient," reveals how far some would go to discredit Kammerer's science. Menninger was not an exception; others echoed his views, especially in Germany and America. American zoologist Howard Parshley, for example, called Kammerer's work pseudoscience and referred to him as a "neo-Lamarckian heretic."[2]

Kammerer, no doubt, had peculiarities. His personality was a maddening combination of vanity and impulsiveness, "bombastishness," a deep insecurity, and an at times extreme emotionality. He was concerned with dress to the point of being a dandy; he was given to dramatic overstatement; and he was sometimes quite depressed. His professional life was replete with near misses and plagued with difficulties. His application for the *Habilitation* (a formal professional certification to teach in German-speaking universities) had been difficult (it was granted in 1910), though in 1909, Kammerer's experimental work on which it was based was awarded an important scientific prize for "pathbreaking physiological discoveries."[3] Yet one of Kammerer's own colleagues had attacked his work; and in the late teens two major German geneticists, Erwin Baur and Max Hartmann, challenged it. Partly on these grounds, in 1919, a proposal to confer on Kammerer the prestigious title of associate professor at the University of Vienna was denied.[4]

Throughout his adult life, Kammerer had polarized his scientific associates, colleagues, and acquaintances. Pioneering British geneticist William Bateson disliked him at first sight and for over a decade challenged his work. After first meeting him in Vienna in 1910, Bateson wrote his wife describing his reaction to the man and his work on the inheritance of acquired characteristics: "there was just the faintest tinge of something like suspicion of humbug in my mind . . . , but taking the whole series of experiments together I cannot really entertain the idea of fraud."[5] He dismissed his doubts as probably due to Kammerer's likeness to Bateson's British competitor, Karl Pearson. But Bateson's assessment became more and more critical, and it was shared by many of his colleagues. In the early 1920s, the premier American geneticist T. H. Morgan refused to be on the scientific committee that welcomed Kammerer on his American lecture tour. In 1923, just before that tour, American biometrician Raymond Pearl told neurologist Henry Donaldson that Kammerer was a fraud. Pearl had been keeping a file on Kammerer since the teens, presumably documenting the grounds for his view. Unfortunately for Pearl, the file was lost in a 1919 fire in his laboratory.[6] Jennings was Pearl's colleague at Johns Hopkins. In his 1924 letter to Harrison endorsing Kammerer, described in Chapter 1, Jennings explained that Kammerer's critics were being unfair, and he implied that Pearl's reaction was especially extreme. He believed that the debate about Kammerer's integrity could be traced to Bateson's influence, which by then was very negative. "It seems to me that he [Bateson] has made a rather sorry spectacle in his attacks: he has continually shifted his ground from one point to another, till from what seemed a serious assault on Kammerer's honesty it has come to a question of certain highly technical details,—with good evidence on Kammerer's side."[7]

But Jennings was unusual. Perhaps as much because of the significance of Kammerer's central concept as the complexity of his personality, few people saw his many sides. Kammerer had spent most of his career working on the controversial concept of the inheritance of acquired characteristics. This was the view, commonly held in the nineteenth century, that change in an individual's actions or surroundings could fundamentally alter heredity, and so be passed on to that individual's offspring. What parents must acquire could become hereditary and automatic in their children. First presented as a scientific view by Jean Baptiste Lamarck in the early nineteenth century, the idea was challenged with the rise of Darwinism. But it reemerged as a force in biology in the 1890s as the "other side" in the debates about nature and nurture that surrounded the birth of modern genetics. By 1920, America led the way in the growth of genetics, but early reforms had also focused on the importance of the environment, and the issue was still discussed. In 1923, the American Philosophical Society held a symposium on the inheritance of acquired characteristics just before Kammerer's arrival in the United States. Despite the dominance of Morgan's work, interest in the concept was present on both sides of the Atlantic.[8]

But by the 1940s, the idea was discarded as a misunderstanding in the early history of genetics. Historians have not, however, fully examined the political and social contexts surrounding the scientific rejection of the idea. Although he notes that debates were rarely decided on the basis of evidence, historian Peter Bowler stresses changes in science, the emergence of genetics, and the separation of evolution from the developmental concepts that dominated evolutionary thinking in the late nineteenth century. Much historical work has also focused on the Nazi period to show how the Nazis aligned themselves with the opposing view for racist and ideological reasons.[9] But little scholarship addresses the social context of the interwar period, when, as the concept lost ground, key transitions were occurring in science and Central Europe was in political turmoil.

Kammerer's story shows that the idea remained a focus of international interest in the 1920s. But historical accounts of his role are dominated by the scandal that occurred in 1926. In that year, an American scientist, one of Kammerer's critics, discovered that some of his zoological specimens had been tampered with. The focus of the fraud was a group of specimens produced in experiments done in 1909, in which Kammerer had placed a land-breeding toad species, the "midwife toad," in water to breed. The male toads developed "nuptial pads," calluses on their paws, produced as they grasped onto females to hold them while they mated in the water. Kammerer showed that over generations, the unexposed descendents of the first generation of water breeders had inherited nuptial pads and the "habit" of breeding in water. Seventeen years later, however, while on a visit to Kammerer's laboratory, American herpetologist G. Kingsley Noble discovered that someone had tampered with the one remaining specimen. India ink had been injected into the pads in an apparent attempt to make them more obvious. No one, including Kammerer, disputed the fraud, which was reported in the pages of the journal *Nature*. Kammerer's boss, Hans Przibram, acknowledged that the data could no longer be used; but he defended Kammerer in print, asserting that the specimens were once valid. Publicly and privately Kammerer denied any role. But the controversy raged, and only months later, in September 1926, while back in Vienna gathering research materials to take to the laboratory at his new position in Moscow, Kammerer shot himself.[10] Except in Austria and Moscow, the suicide was widely interpreted as an admission of guilt.

In his published memorial to Kammerer, Przibram again defended him. Przibram was unconvinced about the validity of the concept; but he separated his conclusions about the idea from his assessment of the strengths and weaknesses of the man. Przibram noted Kammerer's great musical talent, his enormous love of living beings (Przibram had seen none other so great), and his exceptional precision in science. He hinted that Kammerer's unparalleled ability to maintain live animals with such care was one reason why others had been unable to reproduce his work. He also noted that Kammerer's vision of the mental and moral improvement of humanity was controversial and opposed by "eugenicists," who wished to improve the race exclusively through natural selection. Those opposing his ideas fought Kammerer's science not with science, said Przibram, but with suspicions about the experimenter.[11]

Most of the diverse assessments I describe above preceded the discovery of the midwife toad fraud. But the affair confirmed the misgivings and strengthened the case and the will of those disposed to discount Kammerer and his science. Even among supporters like Jennings, the suicide led to a discrediting of his science. But, this was, by and large, not the view of his colleagues or of the Austrian community of scientists. Many rankled at Kammerer's personality, but like Przibram, they nonetheless considered him a talented scientist, an indefatigable worker, and a committed activist and educator. Historian Ludwig Fleck is a striking example of a Central European scholar who, with no apparent vested interest, defended Kammerer's integrity. Fleck's seminal historical work, *The Genesis and Development of a Scientific Fact,* quotes Otto Nägeli's statement that Kammerer's suicide also discredited his early salamander work. Fleck responded with a footnote exonerating Kammerer: "I do not believe in a simple case of bad faith on the part of Kammerer. He was an original and diligent research worker."[12] Fleck was a Polish physician and historian, and we do not know how he knew Kammerer. But his statement, published nine years after Kammerer's suicide, reveals a different view of the scandal voiced by many in Austria. It contrasts strikingly with those of the Germans and Americans who largely dismissed Kammerer's science.

Such extreme reactions often reveal political and social biases on both sides, and this was very likely the case in the many reactions to Kammerer. Who are we to believe, Morgan, Pearl, and Bateson, or Przibram, Jennings, and Fleck? Who really was Paul Kammerer?

Born on August 17, 1880, in Vienna during what author Stefan Zweig calls the "Golden Age of security" and, for Jews, a great age of hope, Kammerer (see fig. 3.1) lived only forty-six years. Though he traveled extensively and lectured throughout Europe, in Russia, and across America, he spent his entire life in the city of Vienna. He was a direct contemporary of Zweig, of the philosopher and reformer Otto Neurath, and of the infamous social critic Otto Weininger. As their works and their fates suggest (three of the four committed suicide), his generation experienced Vienna as a city that would undergo and produce enormous change, some revolutionary and unprecedented, and some tragic.[13]

Kammerer adored the sophisticated culture of the city, and he immersed himself in those changes as they occurred in politics and the arts as well as in science. He came from a middle-class family with a love of music, and even as a boy he developed a passion for living organisms.[14] At home he collected live plants and animals and took special care in maintaining them. In college, he pursued studies in music and philosophy as well as in biology, graduating from both the University of Vienna and the Vienna Music Conservatory. He composed his own songs, and he was active in a circle of musicians that at times included Gustav Mahler, Alban Berg, and Bruno Walter. In 1906, he married the Baroness Felicitas Wiedersperg. They later divorced, but then remarried following the annulment of Kammerer's short marriage to a painter. Though he was said to be very handsome and was considered a ladies' man, Kammerer was also insecure and obsessed about his relations with women. There were many loves, and he often bowed to cultural and scientific heroes toward whom he directed near idolatry. His unusual regard for the composer Mahler may help explain his unrealistic affair with Mahler's famous widow, Alma Mahler-Werfel. Her accounts of his excessive passions give us insight into the man's emotional state. She described him as childlike and naïve. His love letters, she said, made him the "clown of her entire circle." His penchant for hero worship and his histrionic relationships with women paint the picture of exaggerated emotionality in a man who was not content in his accomplishments or his sense of himself.[15]

Kammerer's Career

Kammerer was educated during a time when debates about Darwin and Darwinism dominated biological science. At the turn of the century, "Darwinism" referred not only to the revolutionary theory of natural selection developed by Charles Darwin but also to the broader acceptance of naturalism and

FIGURE 3.1. Paul Kammerer (on far right holding the net) and the Hugo Iltis family catching toads, probably in the early 1920s.

Reproduced with the kind permission of the Hugh Iltis Family.

evolutionary thinking that it had spawned. The concept of natural selection, first published in 1859, convinced the intellectual world of the truth of evolution as a biological process. But many thinkers who accepted evolution disagreed about whether the mechanism of natural selection could work as Darwin had proposed it did. Indeed, in the period between 1890 and 1930, there were many "Darwinisms," some quite different from Darwin's own. Biological science in this period was so well populated with alternative accounts of evolution that Bowler describes the period as one characterized by the "eclipse" of Darwin, that is, of Darwin's mechanism of natural selection.[16] But not of Darwinism. What most of the Darwinisms of the fin de siècle shared was the attempt to account for how descent with modification among related species occurred; how, in the process, plants and animals became adapted to diverse environments; and how, if at all, humanity could be folded into that picture.

Kammerer's young life coincided with the eager reception of Darwinism in German-speaking lands, and because his greatest love was for the magic of animals' abilities to change and adapt, his science was steeped in Darwinism.[17] From 1899 to 1903, among his strongest scientific influences at the University of Vienna were the zoologists Berthold Hatschek and Karl Grobben, and the botanist Richard von Wettstein. In 1902, just before taking a student research trip to the zoological station in Trieste, Kammerer began to work as Przibram's unpaid assistant at the Institute for Experimental Biology. The laboratory was dedicated to a new approach to understanding life termed *Entwicklungsmechanik* (developmental mechanics). Founded in Germany by Wilhelm Roux, Entwicklungsmechanik referred to a way of understanding development that stressed both experimentation and a broad philosophy of science. Przibram's laboratory was revolutionary in implementing the new methodology. Doing so involved moving beyond the comparative and historical studies that had dominated the analysis of embryological development in the nineteenth century and extending methods of controlled experimentation, newly matured in physiology, to the development of animal form and to developmental flexibility.[18] Though technically named the Institute for Experimental Biology, Przibram's lab was popularly called the Vivarium. Both names reflected its philosophy. Most work emphasized the external factors that shaped development and that could be assessed experimentally. Its institutional philosophy avoided "metaphysical" speculation on the impact of unmeasureable internal forces—an effort that had dominated the old embryology. Instead, Vivarium methods often involved maintaining live organisms under controlled conditions to trace their impact for extended periods of time. To many, this extended scientific rigor offered unprecedented promise to understand the mechanical unfolding of development under the influence of outer causal factors.[19]

The Vivarium's facilities were a technical miracle of modern biology. The labs were equipped with a range of modern devices used to house and maintain plants and animals under conditions that permitted the precise control of the conditions shaping their development. At the time, most research animals were just caught in the wild and brought into the artificial conditions of the laboratory. Many survived only briefly; some did not survive at all. But the Vivarium's emphasis on development required that test animals do more than just survive. They had to thrive and to reproduce healthy young. The lab's nickname suggests the novelty of its approach. Animals would not be picked out of ponds and fields, tested, killed, dissected, and discarded. Rather the Institute was a *vivarium*—a facility that sustained life. Animals would be tested as they developed and as they lived, enjoying their natural vitality in captivity. The resulting chronic experiments—like Steinach's—often produced striking and unprecedented results.[20] And it was in maintaining the viability of live animals over generations that Kammerer's talents shone. Just after he announced the opening of the Institute, Przibram engaged Kammerer as a volunteer to maintain the animals housed and studied there, and his skill in their careful nurturing made him central to the Vivarium's mission. He quickly became a paid assistant and worked there until 1924, when he left, frustrated with his situation and his life. But in the early years, the fit could not have been better.

Between 1903 (when he became a regular employee) and 1922, Kammerer's science was devoted to two central themes. Both were linked to adaptation, to evolution, and to development in the German tradition of Darwinism initiated by Ernst Haeckel.[21] The first was his view of symbiosis and cooperation as central processes that worked alongside competition in shaping evolution. The exclusive reliance on competition as a mechanism of evolution troubled him, and he sought to give equal weight to cooperation in producing species change. The second theme, and the one for which he became famous and then infamous, was the inheritance of acquired characteristics. Kammerer attempted to reconcile the new genetics with the concept in a way that stressed the Vivarium's anti-metaphysical, externalist philosophy. The second theme was especially consistent with the Vivarium's emphasis on the impact of external influences on change. And in the early years, the possibility of the inheritance of acquired characteristics was a key theme of Vivarium research.

Before World War I, Kammerer published numerous studies on the problem in reptiles and amphibians. Most examined the role of the environment in producing novel and persistent adaptations. He began his scientific work with a comparison of the reproductive characteristics of two species of salamanders, an alpine and a lowland species. One species, the lowland, laid its eggs in water; the eggs then developed through a series of separate stages in the water environment as the animals gradually became adults. The alpine species, by contrast, normally retained its eggs for a longer period, and individuals developed their adult features not in water, but inside the female. They were born as fully developed adults with no externally living intermediary stages. When Kammerer maintained individuals of the lowland, water-developing species under the dry conditions of land, those conditions lengthened the time that the eggs and larvae remained in the female. Eventually, over generations, in the lowland species too, the young were retained in the female and born as fully developed adults.

This strategy was typical of all of Kammerer's research. Most of his studies were devoted to producing novel adaptations acquired by individuals after exposure to new environmental conditions. Kammerer then kept them alive and continued breeding them to determine whether the changes appeared in later generations, even when the young were raised in the original (unchanged) environment. They usually were; and the effects, Kammerer argued, were due to the induction of novel adaptations by the new environment, which, though acquired in the parents, had become hereditary in their offspring. The direct adjustment of individuals to the new environment had altered the heredity of the species to eventually yield an indirect adaptation in descendents that was hereditary. Older evolutionary attributes were, in this way, replaced by new ones that might not have occurred if natural selection had acted alone. The effects showed, Kammerer believed, that the environment changed heredity, enabling offspring to inherit what their parents must acquire. Just this kind of experiment had produced Jennings's early praise, Bateson's dislike, and Kammerer's final downfall.

In 1926, when Noble discovered that someone had tampered with the remaining midwife toad specimen, Kammerer's enemies implied that he had falsified the evidence. But the specimens were old, originally produced in 1909; they had been through a world war. And as late as 1923, a group of scientists had examined them in Cambridge, England. Several men were unconvinced of the effect, but they noticed nothing unusual. Historian Sander Gliboff's analysis of the fraud suggests that when British scientists examined the specimens, any fraud would have been detected if it had been there from the start.[22] We still do not and probably will never know who injected the ink. But the suicide, as much as the fraud, destroyed Kammerer's reputation and undermined his body of work. In fact, his decline had already begun. He openly acknowledged that his science was affected by his social views; at times he even seemed proud of it. Because he believed that such bias was inevitable, he distrusted those who hid their social concerns behind a patina of objectivity, and he may have been careless with his own. But this was not a widely held view at the time, and at this distance it seems clear that the political implications of his science colored the many reactions to him.

As Kammerer matured, Darwinism became as much as a philosophy of life as a science. For some, that philosophy had hopeful implications for a new natural religion, for education, and for the moral structure of society. Kammerer shared this hope for true moral and cultural exchange between biology and society, a relationship that placed evolutionary concepts at the center of philosophical and ethical hopes for humane social progress. His philosophical studies had placed him within a circle of scholars and activists, mostly men, many socialists, and many Jewish, who wished to reform society on scientific grounds, especially on the basis of Darwinian naturalism. The connection shows that Kammerer's science was not hidden in the laboratory. On the contrary, he was outspokenly committed to public education, to the use of science to promote social reform, and to the application of biology to improve humanity through what he called "organic technology." In popular works and lectures aimed at the public, he regularly touted the social implications of his science, sometimes in passionate tones. Kammerer saw Darwinism both as a biological explanation and as a philosophy of life in which natural processes of evolution should be tightly interwoven with a moral and egalitarian society. His values and his science "were linked not simply by a linear process of application, but by a relationship of reciprocal dependence." His ideas about society required the validity of his science; and his engaged desire for a just society fed his scientific passions.[23] It is likely that he had done reasonable, though not perfect science. But his science trod on very controversial grounds: at its core were a concept and a social framework that he linked to socialism, women's equality, Freemasonry, and the abolition of racial boundaries. And by 1925, the mix had become political fireworks.

Progress, Genetics, and the Inheritance of Acquired Characteristics

In the early nineteenth century, the inheritance of acquired characteristics was associated with Lamarck's pre-Darwinian evolutionary theory. The ascendance of natural selection undermined Lamarckian views of evolution. But the idea that plastic habits (or structures or diseases) acquired directly through use or environmental impact could be passed on and occur automatically in descendents remained prominent. Even Darwin endorsed it, especially in human evolution. In 1910, when Kammerer's work became prominent, many still supported it.[24]

But even as Darwin wrote, the concept was being reshaped. And by 1900, its significance went well beyond scientific biology. Because Darwinism had become the foundation for a natural view of human mind and morality, it raised sensitive social issues that were debated by biologists, psychologists, theologians, reformers, politicians, and philosophers across the political spectrum.

Exclusive reliance on Darwin's mechanism of natural selection implied that the great changes in life's patterns were blind, random, and directionless. For many, the central issue was not fundamentally one of religion (though religion could be a solution), but the question of what in a natural and secular world free of divine creation could guide evolution in moral and progressive ways. What kept life and humanity from evolving willy-nilly in any random (or even brutal) fashion?[25]

Well before Kammerer, two prominent nineteenth-century thinkers turned to the inheritance of acquired characteristics to solve this problem, but first they reconceived the notion for a more thoroughly scientific age. They were Herbert Spencer and Kammerer's idol, Ernst Haeckel. Both sought to embed mind and morality, traditionally the province of religion, in a naturalistic Darwinism. Spencer's account was more social and philosophical than scientific. But it was linked to his thoroughgoing naturalism and to his belief that progress in society follows natural laws just as surely as do the phenomena of physics and life. For Spencer, social laws derive from physical laws in a way that assures the natural foundation of morality. He saw life and mind as one; both obeyed the laws of adaptation by which external environmental conditions produced persistent adjustments in internal organic processes, be they physiology or mind. He proposed a progressive dynamic relationship in which inner attributes rapidly adjusted to outer environmental demands in a perpetual process of "moving equilibration" that depended on the inherited effects of use and disuse—on the inheritance of acquired characteristics, now devoid of Lamarck's evolutionary theory.[26] But although Spencer's views were based in physiology, they were speculative and had little supporting evidence. The inheritance of acquired characteristics was almost a tenet of his speculations. It had to be, or things would not work quite the way he wanted them to work.

The German morphologist Ernst Haeckel had lost his wife Anna in 1864. In his despair, he too, immersed himself in a broad philosophical Darwinism that would restore progress, light, and a sense of direction to life. For Haeckel, as for Spencer, natural direction came from the environment in the form of the inheritance of acquired characteristics. He, too, wished to establish the causal continuity of life and nonlife. If mind and body were one and the laws governing change in life—evolution—were consistent with the natural laws of physics, then intentional purpose and intelligent design had no more place in life and mind than they had in modern physics. Life too, worked mechanistically, like the "clockwork" of a machine. But Haeckel still sought an account that could give change a progressive direction. He found it in a revised concept of the inheritance of acquired characteristics; reworked, it could provide a natural account of how change might be guided by environmentally sensitive principles of flexible heredity: "Every organism transmits to its offspring not only the morphological and physiological characteristics that it inherited from its parents, but also a portion of those characteristics that it acquired itself, during its individual existence, through adaptation."[27] Variable environmental conditions produced infinite variety in the organisms that adjusted to those conditions. If those acquired variants were eventually inherited, they could, in a directional fashion, feed the future process of evolution. Eliminating the assumption (proposed by others) by which organisms purposively adapted to felt needs by *trying* to react to the environment, Haeckel argued instead that environmental change produced bodily variants; an evolutionary adaptation only followed once some variant had been acted on by the body in ways that made it a permanent feature of heredity. By reacting to the environment (and becoming hereditary), the adapted body guided species change.

For those seeking natural progress and moral direction in evolution, Haeckel's view was more palatable than the strong selectionism advocated after 1880 by the ultra-Darwinists, men who promoted natural selection as the only mechanism of evolution. Biology was changing rapidly; genetics and cell biology were becoming significant players in the story of evolution. As early twentieth-century science began to seriously explore cellular structures, such as chromosomes, that hinted at how heredity worked, and as the criteria for evidence stressed experimentation more and more, debates about the validity of the inheritance of acquired characteristics raged. For some, "neo-Lamarckism" seemed to fly in the face of new discoveries about the processes going on within cells. Interpreting the inheritance of acquired characteristics in light of advances in the cellular basis of heredity became a key issue in the modern debates on the problem.

These debates began in earnest at about the time Kammerer was born, when German zoologist/embryologist August Weismann proposed his germ-plasm theory of heredity. Weismann's theory proposed that the germ-plasm was transmitted continuously from parent to offspring; its character was preserved unchanged across generations and not blended or diluted as an organism's features took form. He also asserted that the hereditary material was isolated, sequestered, and protected deep within the nucleus of the cell; it was not distributed through the entirety of the body. This nuclear isolation from other bodily influences helped guarantee the stability of hereditary characteristics—it kept them from willy-nilly changing in any generation, as individuals were exposed to new situations and circumstances. Darwin had made it clear that differential reproduction was the key to evolution conceived as descent with modification, and that barriers to reproduction maintained the evolution of species differences. So stability in the heredity of distinct lines of descent was required both for heredity transmission and for the stable maintenance of species. Weismann's theory proposed how cellular processes might help retain the stable characteristics of individuals, and how, in the process, they kept species stable, too. He interpreted continuity and sequestration in the nucleus to mean that influences originating in the cytoplasm—in the cellular material outside of the cell nucleus—and those originating in somatic cells of the body—all of the cells that were not gametes—could *not* alter the character of the germ-plasm. This was the sticking point. If the germ-plasm were indeed the basis of a continuous and sequestered nuclear heredity, then the inheritance of acquired adaptations was not possible.

Weismann pointedly asserted this implication of his theory. Adaptations acquired through the direct effect of changing conditions could not be inherited. Importantly, however, he did not deny that the environment could affect the germ-plasm. Indeed, well before the idea of mutation assumed its modern form, Weismann proposed that the ultimate source of all variation in the germ-plasm was the environment. He argued, for example, that germ-plasms that were composed of different molecules would react differently to environmental differences in nutrition. This would result in different germ-plasm "growth" and, therefore, in germ variants that could fuel differential inheritance and natural selection. Such conditions could alter the development of the germ-plasm. But Weismann consistently denied that acquired adaptive *outcomes* could in their final form eventually alter the character of the germ-plasm. The environment could simultaneously affect soma and germ in ways that influenced soma development, but persistent adaptive changes in soma (of the kind Haeckel suggested) could not produce separate alterations in the germ-plasm.[28]

Increasingly, as chromosomes were discovered in the cell nucleus and interpreted to be "carriers" of the hereditary material, that material certainly appeared to be continuous and sequestered. After 1900, when attention turned to the meaning of Mendel's laws of hereditary transmission, more and more evidence confirmed that meiosis, the process of cell division that produces sex cells, involved a division of chromosomes in the cell nucleus in which transmissible material was paired before dividing, then halved and preserved, as the father's half and the mother's half reunited in a new individual. Chromosomes reshuffled the material in an orderly process in which "continuous" and sequestered material from the parents was separated and recombined in a variety of possible pairings, each preserving the original parental halves. How, given all this, was the inheritance of characteristics possible? Through what biological process could it occur?

Goal and Purpose in Life

By 1910, many varieties of neo-Lamarckism had emerged; some had little to do with debates about heredity. Another deeper problem lurked in Spencer's and Haeckel's attempts to "rehabilitate" Lamarckism and make it respectable. This was the proper role in biology of teleological explanation, explanation in terms of goals, purpose, and intent. What role did goals and purpose play in an age in which physiology and cell biology had begun to replace design as the cause of life? Much of early nineteenth-century biology had been steeped in vitalism, the scientific view that proposed that the natural laws governing life were different from those guiding lawful regularity in physics and chemistry. The phenomena of physics had no purpose; stones just fell, mechanically, without a goal. Life, however, seemed designed to achieve ends, and early metaphors of adaptation, evolution, and development, including Lamarck's, linked life to immanent processes of forward direction that worked toward ends. That tradition was challenged by the emergence of scientific materialism and by the advances made between 1850 and the 1880s in the burgeoning field of physiology. Many in the so-called physicalist school of biology sought to explain life in terms no different from those used in physics and chemistry, denying purpose and stressing mechanical processes that proceeded, like clockwork, without guidance. After 1859, Darwin's treatment of natural selection became for many part of the physicalist's challenge. Chance and simple subtraction could replace progress in evolution, and the "neo-Darwinism" of Weismann and others strengthened the rejection of assumptions about the role of design and purpose in living nature.

But when natural selection's popularity declined around 1900, some scientists applauded its demise and sought to give progress and purpose a renewed place in biology.[29] Distancing evolution and biology from Lamarck's original teleology had fueled opponents of the inheritance of acquired characteristics. But several of the new Lamarckisms were committed not to eliminating purpose but rather to reconfiguring it and reasserting its importance for a new century of progress. As a result, many varieties of neo-Lamarckism again promoted teleology and the inherently purposive design of biological processes. The most troublesome were those that placed intentional agency within the perceptions or drives of plants and animals. August Pauly's "psycho-Lamarckism," for example, reintroduced concepts of mind into evolution through embryology, holding that conscious choice was involved in the origin of organic form.[30] As Weismann's challenge solidified after the rediscovery of Mendel's laws, the validity of the inheritance of acquired characteristics was once again intertwined with debates about the inherent purposefulness of life.

The idea that the inheritance of acquired characteristics might still provide a solution to the problem of purpose especially captured Viennese thinkers. In 1902, the Vienna Philosophical Society published a series of articles on the "crisis of Darwinism." Several authors (including Kammerer's zoology professor) defended neo-Lamarckism on the then familiar grounds that selection was an implausible explanation for the evolution of very complex evolutionary adaptations. Chance and vast amounts of time might make them possible, but they also rendered them very unlikely. Could, as Haeckel had suggested, a reassertion of the inheritance of acquired characteristics solve the problem? One contribution in the series addressed the deeper philosophical issue surrounding the several neo-Lamarckian approaches discussed: the issue of teleology in a post-selectionist biology. It was written by the Viennese physician, Freud collaborator, and respected sensory physiologist Josef Breuer.

Several speakers at the conference had argued that neo-Lamarckian alternatives could replace the "bankrupt" approach that emphasized only natural selection. Breuer did not challenge them; he focused instead on the fact that much more was at stake than just rejecting a theory. Breuer feared that the renunciation of selection might lead to the return of a theologically based teleological worldview.[31] He argued that rampant teleology was inappropriate in a scientific and post-Kantian biology. After Darwin, biologists understood that the organic world "is the last link in an immense chain of transformations, an infinite Becoming." Understanding causation in such a chain required appreciating that immediate conditions caused biological change, not goals. But was that the whole story? Science sought "causal interconnections" among conditions. But the rejection of purpose *and* selection left a major gap: without one or the other, the problem of utility (*Zweckmässigkeit*) remained unresolved. Establishing the relationship between an end and some kind of path to that end was indeed, Breuer said, a problem that biology must address. Without it there could be no account of how evolution "favors the useful and goal-directed (*nützlichen, zweckmäßigen*) path over another."[32] Breuer seemed dissatisfied with Lamarckian alternatives; and he saw selection as a shaper of the organic world. But it was only a corrective; "the driving force [of evolution] must have its roots elsewhere." He offered no solution to the problem of usefulness. But his deliberations illustrate how prominent the debates about teleology were in Vienna, and how important neo-Lamarckian versus selectionist views of heredity were in these debates. At issue was not just flexibility in heredity but also an appropriate philosophical framework for explanation in modern biology, one that in a post-Kantian world could offer a middle ground between intentional design and pure chance.

Physiological Solutions

Haeckel had championed external causes over what he regarded as the metaphysical and religious mythology that crept into biology with speculation on internal causes. This led him to reject the internalist "metaphysics" of Weismann. But one could not reject the problem. If the environment were that powerful, how could nature achieve stability? The key was to determine the means by which the effects of the environment could sometimes penetrate the isolation of the nucleus and induce changes in heredity. Physiology had led the way in the original challenge to vitalism, and Haeckel saw physiology as the answer here, too. So it is not surprising that one of the most comprehensive attempts to develop a physiological account of the inheritance of acquired characteristics before Kammerer was proposed by one of Haeckel's most promising students, Richard Semon. Semon, too, became Kammerer's hero, and in him we see the link between Haeckel's account of the inheritance of acquired characteristics and Kammerer's recognition of the need for a physiological explanation of the process.

After surviving malaria following a failed research expedition to Africa, Semon returned to the University of Jena as Haeckel's assistant. There he completed important work on the embryonic development of Australian lungfish and was named an associate professor. But Semon's career was ruined when, in 1897, he had an affair and eloped with the wife of a Jena colleague. He was fired in the wake of the ensuing scandal, and the two lovers moved to Munich. There, with no laboratory in which to work, Semon focused on developing a theory of the bodily mechanisms underlying the inheritance of acquired characteristics: he called it the mneme theory.[33] The theory drew on a broad organic concept of memory based in physical traces of environmental events, which Semon called "engrams." They were not conscious cognitive or neural traces, but the pervasive physiological remnants of external stimulation that were transferred from cell to cell. But, though he continued to cite the experiments of Kammerer and others, Semon was no longer in a position to pursue the idea empirically. In 1918, despondent over the death of his wife and demoralized by the war, Semon committed suicide.

Semon's theory was speculative and controversial, and he publicly took on the exclusiveness of Weismann's view. Semon also expressed great respect for Weismann's effort; he had extended the selection theory and moved the debate about use inheritance from the realm of anecdote into that of science. But Weismann had gone too far in altogether disallowing the possibility that environment might alter the germ-plasm. Why should science not seek evidence on the conditions in which nuclear and chromosomal barriers could be crossed? Semon argued that Weismann's account failed to explain the available evidence, and much of his thinking drew on a distinction made by the Jena botanist Carl Detto.

In a book published in 1904, Detto described two different kinds of environmental effects on heredity: parallel induction and somatic induction. In parallel induction, the effect of the environment on the germ-plasm was direct and pervasive: environmental change altered the body at the same time as it altered the germ-plasm, changing germ and soma in parallel. And the changes were general, like malnutrition; they usually influenced the whole body. But no prior adjustment in the body had altered the germ; instead each reacted to a common influence. In what Detto called somatic induction, on the other hand, the environment first altered the soma in a persistent and specific way: an environmental change produced an acquired, nonhereditary adaptation (or maladaptation) in some feature of an individual's body. An organ adjusted—adapted—in direct response to the environment, but with no immediate change in heredity. Sometimes the persistence of the effect on soma then also triggered a process that altered the germ. In somatic induction, therefore, any effect on heredity was an indirect one mediated by a prior *functional* change in the soma that was specific to an organ or system: environmental change first altered the soma, and that somatic change eventually triggered a separate physiologically induced change in heredity.[34]

Framed in terms of somatic induction, physiology could potentially account for the manner in which somatic adjustment might penetrate the nucleus to alter the production of the germ-plasm. If proponents of the inheritance of acquired characteristics could explain how physiology mediated an environmentally induced change in germ, they might prevail. Semon had tried to rule out parallel induction (which Weismann accepted) and instead interpret the results of breeding experiments showing the inheritance of acquired characteristics in terms of "engrams" produced by residual stimulation (*Reizübertragung*) that had triggered somatic induction. But in an age that increasingly valued the experiment, his account of how it all occurred was too speculative. There was no evidence for it, and it smacked of psychic processes.[35]

Detto's analysis more clearly broached the question of evidence. He dealt extensively with two problems that consistently reappeared in the debate: the rejection of teleology in accounts of the utility of developing attributes, and a set of empirical criteria that distinguished somatic from parallel induction and, therefore, could be used to conclude that one or the other might account for a particular case of hereditary change. His analysis shows two important things. First, rejecting the inheritance of acquired characteristics was a philosophically important aspect of Weismannist predeterminism. Its value touched on arguably the most important issue facing biology at the turn of the twentieth century: the debate about physicalist causes versus goals. Second, Detto outlined empirical alternatives in physiology that could make environmental control of heredity consistent with physicalism. For ardent selectionists who rejected somatic induction, linking utility to structure without teleology rested with the assumption of germ-plasm–based predeterminants—today's genes—that mechanically dictated adaptive results. But Detto's criteria presented physiological mediation as a third alternative that was neither teleological nor predetermined, and could be anchored in physical causes.

Detto agreed with Weismann's account of parallel induction; adaptive change sometimes arose, indirectly in response to a prior environmental change after selection acted on hereditary structures altered by the environment. But he also took a major step away from Weismann, who had argued that effects of the environment had "nothing to do with the view that functional modifications of any particular organ can cause a corresponding change in the germ-plasm."[36] Sometimes, Detto said, functional modifications could produce somatic induction. Nothing in principle prevented it, and he sketched various conditions that might help establish how somatic induction could occur. He described a number of possible physiological influences that might be involved, including newly produced somatic substances, or "metagenetic secretions." These were novel cell products created by environmental stimulation that might be transferred to the germ-plasm by a kind of "conducting stimulation (*Leitungsreize*)." He included no real evidence. Rather, his point was that nothing in principle prevented somatic cell products from transforming germ cells. All that was needed was to apply the criteria set forth and work out the details of the relevant physiology.

Detto's analysis set the terms of the debate for several decades. His sketches offered a scientifically respectable alternative for those, like Kammerer, Haeckel, and Semon, who were suspicious of hereditary predeterminism but sought causal-physiological influences on heredity that were based in physical processes. Kammerer and Semon both drew on his distinction and presented their accounts of flexible heredity in terms of somatic induction. As early as 1911, Kammerer had relied on the concept to interpret his zoological demonstrations of the inheritance of acquired characteristics. Referring to alternative accounts offered by critics, he wrote: "These possible explanations still seem to me . . . to be somewhat far-fetched and decidedly more contrived than the assumption of somatic induction, an assumption which there is no longer any sufficient reason to avoid."[37]

Increasingly, however, it could not remain an assumption. The lack of evidence on the mechanism of somatic induction was one reason why American neo-Lamarckians, such as the American botanist Daniel MacDougal, were skeptical of Kammerer's zoological results. Speaking for the scientific community in 1924, MacDougal wrote Kammerer: "we have not yet found a way in which the experience of individuals may be impressed permanently upon the germ-plasm. It is my opinion that in finding this out we must uncover some new relations in physiology, or determine the stage or condition of the soma and of the germ-plasm in which this may occur."[38]

Historian Jonathan Harwood has described important differences in how quickly Weismann's theory was accepted in Central European as against American genetics. American geneticists accepted his ideas quickly, especially after T. H. Morgan's *Drosophila* school came to dominate American genetics. But Central European scientists remained more skeptical. By the 1920s, often with good cellular evidence, several challenged the assumption of nuclear sequestration using, for example, the concept of cytoplasmic inheritance, a view that advocated some other kind of hereditary unit or structure that existed outside of the cell nucleus (in the cytoplasm) and that shaped organisms independently of or complementary to genes.[39] They, too, often looked to physiology to explain the impact of extranuclear structures or processes, and Kammerer often cited them. So in searching for alternative views of heredity that looked outside the nucleus, Kammerer was by no means alone. Physiological solutions were in the air, at least in Central Europe. Kammerer's early work had demonstrated that effects acquired in one generation could become hereditary in the next; but he could not rule out parallel induction. Detto had given him a key: individuals' adjustments to their immediate environments were possible when mediated by "metagenetic secretions." Perhaps with this in mind, he turned to endocrinology to pursue how novel adaptive outcomes could be induced in a flexible system that depended, not on goals or hereditary predetermination but on somatic adjustments that reached the germ-plasm via hormones.

The Gonads and Vivarium Externalism

In 1911, Vivarium Director Hans Przibram published a paper outlining a program of research that was to guide Vivarium research for more than a decade. In it he said, "The question of the inheritance of acquired characteristics has entered a new stage."[40] To describe why, he cited Semon's challenge to Weismann's argument that all environmental effects on heredity were due to parallel induction. Przibram was not fully committed to the inheritance of acquired characteristics; the evidence was too scanty, the idea still too theoretical, and he remained skeptical. But he believed that an empirical assault on the question was the next step in settling the issue, one important enough to become central to the Vivarium's research program. His goal was to define the details of that program. At issue was not whether characters could be produced in parents that persisted in offspring—that had been repeatedly demonstrated. The problem was how, by what physiological mechanism, were the changes in offspring brought about? Were they due to somatic induction, and, if so, how did it occur?

The paper was the first in a series of Vivarium research reports that all began with the title "Die Umwelt des Keimplasmas" (The environment of the germ-plasm). Each paper in the series stressed the impact of external influences on development. In the first, Przibram focused on an empirical program that could help confirm the mechanism of somatic induction. The key to that program, Przibram said, was the gonad. The first step in explaining somatic induction would be to understand the gonad as the immediate local environment of the germ-plasm. He identified three important research areas that would help do this: scientists must examine the physical conditions in which the gonads normally exist and which might influence their products directly; they must then trace changes in the gonads that could result from changes in the external physical environment; and finally, they must examine the dynamic interrelationships (*Wechselbeziehungen*) between the gonads and the rest of the body, "between their internal workings (*Binnenwelt*) and their environment in the narrow sense of the word."[41] Gonads had already been the focus of much inquiry in the study of heredity. The organs, after all, produced the germ-plasm. Przibram's plan, however, stressed that the gonads, too, existed in environments. As he described each step of the plan, he reiterated the importance of understanding the inner world of the gonads and of examining how their functions might be modified by other systems in the body, by the soma. The morphology, secretions, and sensitivities of the gonads were the physiological environment of the germ cells. They were, therefore, also the likely place to look to extend physiology to the problem of somatic induction.

Przibram did not raise the double gonad question, but he implied it by framing somatic induction in terms of the internal environment of the gonad. If the germ glands were only half of a double gland, and if the germinal half were bathed in secretions from the other half in a way that affected generative cell development, the secretions of the interstitial cells would be a central dimension of that internal world. Tandler and Steinach had both already shown that, interpreted as a separate organ, interstitial cells could alter germ cell production. Przibram's program implied that more information was needed. To understand the inner world of the gonad, to relate that knowledge to other bodily systems and to their reaction to changes in the environment, the Vivarium needed a reproductive endocrinologist committed to the interstitial cells. In the same year the paper appeared, Przibram was in the process of getting one; he had just hired Steinach to head the Vivarium's Physiology Department.

Steinach arrived in 1912, just as his demonstrations of sexual development were beginning to appear in the scientific literature. By 1916, his work on experimental hermaphrodites had demonstrated that male and female characteristics could develop in the same individual. And by 1918, there was considerable evidence indicating that the interstitial cells constituted a separate gland that secreted the hormones controlling sexual development. Kammerer took special notice. He had speculated on sexuality and sexual physiology from an early stage in his career. But his research had concentrated on morphological changes—not hormones—in salamanders, frogs, and toads. He had praised Semon and relied on his theory to assume the physiological foundation of somatic induction.[42] Now Kammerer's boss established the environment of the gonad as a lab priority and gave him a colleague with the expertise to explore it. By the late teens, Kammerer began to take a radically new approach to settling the debates surrounding his life's passion: the inheritance of acquired characteristics. He aligned himself with Steinach and jumped onto the interstitial cell hypothesis bandwagon. But Kammerer was not an endocrinologist, and sexual reform was not the issue he sought to pursue. Rather, using a concept already steeped in controversies over the roles of men and women in society, Kammerer took on the equally contentious issue of the role of the double gonad in heredity. He aligned himself with Steinach and the interstitial cells to tackle the mechanism of somatic induction. But in doing so, the troubled Kammerer inherited a whole new set of problems, conceptual and personal.

The Kammerer-Steinach Relationship

Steinach, too, was immersed in scientific controversy that was fed by complex social issues. He had, after all, chemically fused masculinity and femininity in the same individuals. In the social turmoil that engulfed gender roles after World War I, his work on the malleability of mammalian sexual development was inflammatory (see Chapter 2). Steinach fueled the fire by attempting to interpret homosexuality, not as a degenerative disease but as a natural product of endocrine development. And as if that were not enough, in 1920, at the beginning of a controversy that deeply touched the United States, Steinach announced that he had developed a procedure for bilateral vasectomy that could fend off aging, rejuvenate the elderly, and help restore the ailing body to youthful vigor. It did so by increasing the number of interstitial cells in the gonad. Their elevated secretions, Steinach believed, stimulated a bodily reinvigoration, which he termed "rejuvenation (*Verjungung*)."

The idea was controversial on both sides of the Atlantic, and it became so closely linked to Steinach that the procedure was popularly called "being Steinached." Rejuvenation, the separation of sexuality from reproduction, bisexuality, even the secretions themselves—each was linked to scientific disputes in which Steinach and the interstitial cells were embroiled.[43] Nonetheless, Kammerer eagerly aligned himself with Steinach's work and with the interstitial cells. His American lecture tours vigorously promoted rejuvenation to stress the curative effects of hormones. The controversies surrounding Steinach probably clouded the reception of Kammerer's endocrine hypothesis of somatic induction. But beyond the reactions to his work, there were personal problems. Steinach was a man of explosive fits. He was obsessed with his scientific reputation, often unjustly berating his colleagues, even those, like Kammerer, who worked devotedly to promote his research. In becoming a defender of Steinach, Kammerer had stepped into a fresh minefield.

In 1920, Kammerer and Steinach independently began corresponding with the German American physician Harry Benjamin. Benjamin was an endocrinologist and a pioneer in understanding transsexuality. A recent U.S. immigrant, after the war he traveled to Europe and visited the Vivarium. He quickly became convinced of the value of Steinach's research and his rejuvenation procedure. After returning to New York, Benjamin spent the next twenty years trying to convince American physicians that Steinach's surgical procedures could indeed allay the aging process by stimulating hormone production. He gathered data on the health benefits of the procedure, he promoted its value to the public, he lobbied to establish a Steinach institute in New York, and, he tried to help Steinach immigrate to the United States when Steinach was marooned in Switzerland in 1938.

Steinach's reaction to Benjamin's efforts illustrates how he dealt even with his most supportive colleagues. The two men disagreed on matters of public exposure. Benjamin believed that American physicians would never see aging as a disease, as something that needed to be treated. Popular acceptance of rejuvenation, he thought, depended on a publicity campaign that bypassed physicians and went directly to the people via advertising. Only if people asked their doctors to perform the procedure would physicians actually use it. But Steinach mistrusted advertising, especially in the American press. He feared that oversimplification and wide exposure would distort the quality of his science and leave him in a cloud of sensationalism. Kammerer and Benjamin thought the fear exaggerated. But in part, Steinach was right—popular attention did distort rejuvenation's claims and its promise.[44] In response to the concerns, however, Steinach regularly attacked Benjamin, accusing him in abusive and near-hysterical outbursts of trying to undermine Steinach's work. The tone of his mistrust is palpable. Benjamin almost always acted in Steinach's best interests. But to pursue them he had to regularly assuage the Viennese endocrinologist, stroking him, and reminding Steinach of his assessment of Steinach's greatness.[45]

Kammerer met Benjamin during one of the latter's first visits to the Vivarium. The two men became very close friends and corresponded until Kammerer's death. They shared intimate details of their lives—Kammerer, for example, revealed his depression and spoke of several suicide attempts, though not the last one. Their mutual trust deepened to the extent that Kammerer gave Benjamin his power of attorney, authorizing Benjamin to handle his legal and financial affairs in the United States. When Kammerer's job prospects in Austria left him financially desperate, Benjamin tried to find him teaching positions in America. To help, Benjamin sold Kammerer's books to U.S. colleagues, returning the money to Kammerer. The two men also spoke of Steinach, and their private thoughts reveal how much Kammerer had invested in the Kammerer-Steinach relationship. For Kammerer too, Steinach was a difficult man, given to fits of rage and crafting accusatory scenes of scientific persecution. In 1922, Kammerer wrote Benjamin: "The path is not strewn with roses, but the work gives me great joy and is already successful."[46] The first phrase was a rare understatement. In December of that year, seeing how sensitive Steinach was, Benjamin advised Kammerer, "It is best that you do *not* speak of our correspondence to Steinach. . . . You know how mistrustful he is."[47]

As Kammerer grew to see the importance of a physiological account of the inheritance of acquired characteristics, he sought research space in Steinach's laboratory. Beyond that involved in their one collaboration (see Chapter 4), he never received it. A photograph in Steinach's scientific autobiography, *Sex and Life,* shows Kammerer in a picture that portrays "Steinach among his students" (see fig. 3.2).[48] But Kammerer was not Steinach's student, and Steinach was ambivalent about Kammerer's role. Embroiled in his own controversies about the interstitial cells, the rejuvenation debate, and his interpretation of homosexuality, Steinach wished to avoid linking himself to the controversies that surrounded Kammerer. Kammerer saw Steinach's reluctance. In a letter to Heinrich Glücksmann, Kammerer quoted the response he had received from Steinach after forwarding an inquiry from Glücksmann about lab space for some Dutch physicians. Steinach wrote: "For the moment my laboratory exists only on paper and nothing can be done. As soon as my new laboratory is set up—with pleasure."[49] Steinach presented himself as enthusiastic, but there were no new facilities and the prospects were poor, given the postwar financial circumstances. And Kammerer knew that Steinach's lab existed more than just "on paper."[50] He told Glücksmann that he had predicted Steinach's evasion. He found so many meanings in Steinach's response that he asked Glücksmann to help decode it. The answer is lost, but the question shows that he didn't accept Steinach's response at face value. The brush-off was motivated in part by Steinach's fear that he would be tainted by Kammerer's reputation. He said as much in a letter to Benjamin in 1923. Kammerer's advertisements to promote him were, Steinach said, "'for me'—that means: 'against me.'" He recognized the value of Kammerer's book praising rejuvenation. But it was a popular summary, and Steinach denied any deeper scientific involvement. Beyond their one collaboration, he said, Kammerer's contribution was only literary. "Kammerer is *not* . . . my coworker (*Mitarbeiter*)."[51]

Steinach considered Kammerer and Benjamin potential competitors, even though neither could claim a scientific reputation anywhere near that of Steinach, and neither really tried. One example, a serious break, illustrates the tensions that existed among them. In 1923, the *New York Times* published an article summarizing Kammerer's work, exaggerating its significance, and mistakenly attributing much of Steinach's research on rejuvenation to Kammerer.[52] Steinach received a copy of the article from Simon Flexner, then director of the Rockefeller Institute. He was outraged.[53] In the aftermath, Kammerer tried to correct the error. He wrote Steinach to apologize and convince him that he had never

FIGURE 3.2. Eugen Steinach's laboratory personnel, showing Kammerer as one of Steinach's students. Kammerer is on the far left (rear), Steinach in the center (with the beard).

From Eugen Steinach, *Sex and Life: Forty Years of Biological and Medical Experiments* (New York: Viking, 1940), 188.

presented himself as a force behind Steinach's work. The letter is one of few communications known between the two men. It shows Kammerer's emotional investment in the connection with endocrinology, his fear for his future, and his great sense of disappointment that Steinach had questioned his honor. Kammerer's self-presentation combines a palpable dejection tinged with martyrdom and desperation. In it, he said that he had played no part in arranging the *Times* interview, and he never claimed originality for Steinach's work. He also confessed great disappointment that Steinach could conclude that he had tried. He asked Steinach to trust his complete integrity, both personally and professionally, and to understand his situation. The Vivarium was no longer paying him; all of his efforts to support Steinach's work had been driven by his devotion to Steinach, his ideas, and the amazing possibilities they offered. His chances in America, Kammerer wrote, were his last hope to avoid becoming a beggar and prostituting himself in an effort to survive. He wrote, "The only thing left to me would have been the *honor* (sometimes justifiably discomforted on account of the combined attacks), to be called your associate, that is to say, to be permitted to be recognized as such by you. . . . The craftsman (I was . . . nothing more than that) does not ask for that, but must be paid according to the agreed upon *value* (not money). The present situation offers the opportunity to [pay] through these values and replace my heavily felt material loss elsewhere. I had hoped never to have to mention this connection or at all the heavy losses that I suffered by working on the film, etc."[54] Being his associate would have sufficed; but, because Steinach had so mistrusted his motives, he was reduced to requesting monetary payment. He had received little gratitude; and worse, his honor had been questioned by the very one to whom he had devoted himself and on whom his future depended. He had risked his personal and his professional future on the volatile endocrinologist, and this was his thanks.[55]

But Kammerer's martyrdom minimized his role in the flack. Kammerer knew of Steinach's fear of advertising—he and Benjamin had discussed it. But he nonetheless flaunted his connection, aggravating what he knew could be a problem. Advertisements for his rejuvenation book, which were almost surely written by Kammerer or Benjamin, stated that the book was "Written by a great scientist who has been called 'the collaborator, friend, and interpreter' of Dr. Steinach."[56] Collaboration with Steinach was indeed no bed of roses. But Kammerer's self-promotion did little to avoid the tensions. Instead, amid debates about mechanism and teleology, intersexuality, and rejuvenation, Kammerer strutted the change in his scientific direction, one made possible only by dealing with the difficult Eugen Steinach. Why did he persist?

Semon and Detto had placed the problem of somatic induction squarely within the domain of physiology; Przibram made it central to the Vivarium's research program and lodged the question, anatomically and physiologically, in the gonad. But none of them had the empirical evidence needed to settle the disputes about somatic induction. In hiring Steinach, Przibram gave Kammerer a colleague with the expertise to attack the question. Kammerer saw that by appeasing his temperamental colleague he could draw on the expertise of a scientist whose approach offered an empirical solution to the problem. But the switch was not based solely in the pragmatic need for evidence. Steinach's science had indeed convinced Kammerer of the value of the new approach. But Kammerer, the utopian visionary, had other motives, motives involving revision of the scientific concept of race.

4

Sex, Race, and Heat Rats

Somatic Induction and the Double Gonad

Paul Kammerer had amassed good zoological evidence for the inheritance of acquired characteristics. This was one reason why he was at the center of so much controversy. But new cellular evidence on the mechanisms of heredity—the discovery of chromosomes and the stability and reappearance of attributes in hereditary transmission—raised important scientific questions that required a different kind of answer. The validity of the idea no longer depended on zoological demonstrations like the ones that Kammerer had developed. Instead, evidence in physiology and cell biology was necessary to bring the concept into modern discussions framed by chromosomes and the process of meiosis. Semon's theories had focused the issue, but in a language that did not produce meaningful scientific tests. Indeed, until Przibram's 1911 call to examine the gonad, most appeals to physiology were vague and speculative.

After Przibram hired Steinach and evidence on the interstitial cell question had begun to favor a secretory function, Kammerer could avail himself of a valuable new approach. All of Steinach's research promoted the idea that the gonad was a double gland—one structure that functioned as two separate glands, one serving reproduction and the other sexuality proper (see Chapter 2). In 1915, many scientists still treated the two functions as one and vested them in eggs and sperm. But Steinach's rise was based on his confirmation of two glands with quite different functions, and by 1919, Kammerer had become a vocal supporter of Steinach's research on the interstitial cells and flexibility in sexual development. He began to argue that the double gonad could also provide the cellular mechanism needed to explain how the inheritance of acquired characteristics occurred. The interstitial cells were the key to somatic induction, and collaboration with Steinach would yield the needed evidence. But Kammerer's motives were also social. After the war ended, he became more concerned about the growing rigidity of the race concept in science. He believed that if research on the mammalian double gonad could solve the problem of somatic induction, it could help explain racial differences in human sexuality flexibly without recourse to rigid racial boundaries.

Steinach and Kammerer's experiments focused on rat sexual development. But they extended their collaboration well beyond the animal data to address the sex differences described among human populations. Anthropologists had distinguished the so-called "civilized" peoples of Europe, then called *Kulturvölker*, from what many regarded as the "primitive" peoples of Africa, native America, and the Polynesian Islands, whom they called *Naturvölker*. The terminology reflects their assumption that civilization, "culture," and the proper sexual restraint and specialization they entailed had evolved only in European "races." "Races" of natural people were assumed to be more like animals. Kammerer disagreed; his goal was to account for differences in the presumed racial sexuality of humans by rejecting the rigidity of the race concept, a concept that by 1920 was anchored in Weismannist genetics.[1] Instead, he and Steinach would rely on inferences from their rat tests to show that racial differences in human sexuality were the result of climate acting on heredity in a way mediated by somatic induction.

The Endocrine Hypothesis of Somatic Induction

Paul Kammerer's interests in sexuality and physiology were long-standing. In the program of study that he proposed in his 1908 application for the *Habilitation*, Kammerer listed the comparative physiology of reproduction and the physiology of heredity among the several fields that he would master.[2] In 1912, he compiled that information into a comprehensive review of the question of sex differences that was published in Emil Abderhalden's influential *Fortschritte der naturwissenschaftliche Forschung* (Progress in scientific research).[3] The review was impressively comprehensive; it treated the science of reproduction as a far-reaching window into the problems of species and racial development and the nature of heredity. Kammerer dealt with confusions in the common terminology used to describe sex differences (primary, secondary, and tertiary), with sex ratios, with Morgan and Edmund Wilson's heterochromosome theory, and with Richard Hertwig's ideas on the relationship between cytoplasm and germ-plasm—hinting that chromosomes could incorporate aspects of the outer environment. He reviewed several ideas about the origin of sex differences, and supported the new interstitial cell hypothesis of Ancel, Bouin, and others. Steinach was mentioned, but only as a minor figure. Kammerer placed greater emphasis on the work of Tandler, Hertwig, Moritz Nußbaum, and others. Most of Steinach's transplant work had not yet been done; in 1912, he was only one among several who were exploring the hypothesis.

Things had changed a great deal by 1919, when Kammerer's second review appeared: Steinach was its central focus. Separated by only seven years, the differences between the papers show the power of Kammerer's conversion to Steinach's endocrinology. Kammerer alluded to the inheritance of acquired characteristics in the first review, but only in functional terms, referring to a cell's ability to chemically assimilate environmental influences.[4] By 1919, however, his conceptual emphasis had shifted to the mechanism of somatic induction, and he was hard on the trail of an endocrine solution. The key was the validity of the double gonad, especially as framed by the evolutionary and embryological relationship between its two parts. How did they interact in ways that shaped the development of embryonic sexual form?[5]

The stated intent of Kammerer's 1919 review was not simply to summarize Steinach's findings, but to lay out their significance for broader issues in biology. He presented himself as an advocate, providing a comprehensive account of Steinach's work, which (compared to the brief mention in 1912) even included the details of experimental controls. Throughout, following Steinach, Kammerer stressed the impact of hormones on the central nervous system. The blood-borne chemicals produced by the interstitial cells worked through the nervous system—Steinach's erotization—to cause "psychic" sexuality: desires, mental masculinity and femininity, and even homosexuality.

Kammerer presented many details of cellular endocrinology, including the work of Sand, Lipschütz, Tandler, and Artur Foges, all of whose data largely confirmed Steinach's conclusion. He summarized Tandler's work on cryptorchidism (the failure of the testicles to descend to the scrotum), and the collaboration between Steinach and the radiologist Guido Holzknecht, which showed that exposure of infant female mammals to x-rays exaggerated the development of the mammary glands and uterus and produced huge increases in interstitial cells. Kammerer wrote that Steinach's work had convinced many that "only interstitial tissues could be responsible for the development of somatic and mental sexual attributes."[6] But the force of the paper depended on how Kammerer broadened Steinach's approach to place the question of the double gonad in a phylogenetic and developmental context. He repositioned what had been a question of sexual anatomy and physiology within the purview of the evolution of sexual differentiation. While others had engaged in similar exercises,[7] Kammerer followed Tandler and Grosz in stressing the evolutionary and developmental implications of the gonad as two glands that separated the function of reproduction from that of sexuality. The task was great. There were large numbers of contradictory findings that could be interpreted either as failures to replicate or as true evolutionary differences among species. Kammerer had to build a framework that could disentangle methodological differences from biological ones—many species did not share the mammalian glandular structure—and that could reflect greater or lesser specialization in the evolution of the endocrine bases of sexuality.

Kammerer spent several pages addressing Bidder's organ in toads. His attempt illustrates both the problem he faced and his solution. Bidder's organ is a structure near the testes of male toads. At the time it was considered a rudimentary ovary. Though it appeared to contain no interstitial cells, research had shown that when the organ was retransplanted into male castrates, the animals developed secondary sexual attributes. Castrated males with implanted Bidder's organs even developed functioning ovaries. To many, this meant that the interstitial cells were not necessary for the development of secondary sexual characteristics. But others had described a dark secretion in the degenerated ova of Bidder's organ. Kammerer took this to mean that the organ was a secretory structure; and he suggested that it might reflect an incompletely differentiated double gonad. Phylogenetically, mammalian double gonads were highly specialized structures; toads' Bidder's organs—structures not yet fully specialized for secretion—were evidence of an early evolutionary stage prior to the full transition to the mammalian specialization. Throughout the review, Kammerer drew similar inferences integrating evolutionary and developmental perspectives on interstitial cell differentiation with endocrine data gathered on mammals. No inconsistency was introduced by the lack of interstitial cells in some species of insects and amphibians: there, like the toads, the evolutionary transitions to the specialization seen in mammals were either incomplete or of no use. Kammerer wrote: "In vertebrates the differentiation is complete. But we also encounter intermediary cases in which the endocrine task is allocated to a distinct rudimentary gonad (Bidder's organ) that contains no interstitial tissue. Rather, here the reproductive function of the germinal tissue relinquishes the aid of the hormonal function. In mammals, where only Steinach has established and claimed the hormonal impact of the interstitial cells, the highpoint of this specialization has been reached."[8] With his evolutionary perspective Kammerer turned disadvantage into advantage: invertebrates and toads were different because they had not yet attained the specialization that was clear in the two distinct glands of mammals. What some perceived as a problem was in fact an evolutionary difference confirming the role of the interstitial cells as a specialized mammalian sexual adaptation.

His evolutionary and morphogenetic perspective led Kammerer to a radical new hypothesis. This was the basis of his interest in integrating Steinach's endocrinology with the phylogeny of cellular differentiation in gonads. Kammerer argued that the evidence he assembled implied that interstitial cells and generative cells (eggs and sperm) had in evolution derived from the same primitive germ cells (*Urgeschlechtszellen*). In many vertebrates, the two cell types had evolved distinct and specialized secretory functions: one "secreted" gametes;

the other secreted the hormones that shaped sexual attributes. In mammals, they were morphologically distinct. But physiologically and evolutionarily, they were connected: they engaged in an "intimate and dynamic developmental relationship (*wechselseitige genetische Beziehung*)." And because of their morphogenetic connection, the hormonal specialization of mammalian interstitial cells was still closely tied to the germinal function of gamete production.[9]

Kammerer's speculations are rarely tightly connected in the paper. But his several converging perspectives generated a framework that allowed him to stress the common heritage of the two cell types. The motivation for this claim was not clear in the 1919 paper. But it became clear in his collaboration with Steinach, an effort that probably began in the year the review appeared. Steinach and Kammerer's research provided the first experimental evidence supporting the idea that the "intimate developmental relationship" of the two cell types could solve the problem of somatic induction.

Heat Rats, Race, and Human Sexuality

The Steinach-Kammerer collaboration is a very wide-ranging effort, even by the standards of the time. The work is a more than sixty-page paper that was published in one of the most important biology journals of the day. On the surface, the paper simply reports a series of experiments done on rodents—"heat rats"—which had been raised in unusually high temperatures. But in it the authors ranged from rats to race, and to the flexible environmental basis for racial differences in human sexuality. Their title focused on the central human question: "Climate and Puberty (*Klima und Mannbarkeit*),"[10] and they argued that human racial differences were due to heat, to the torrid tropical climates in which many Naturvölker lived. They based their treatment on Steinach's conclusion that the interstitial cells were the mammalian puberty gland. Steinach had shown repeatedly that the extent of sexual development, sexual maturity, and the strength of mammals' sex drives were directly proportional to the amount of interstitial tissue in the gonads. Even natural fluctuations in individual sexual maturity and precocious puberty were accompanied by differences in the interstitial cells. The authors' goal was to show that these changes were not strictly intrinsic; rather, they could occur when individuals moved from a cool climate to a hot one. Almost certainly responding to Kammerer's central focus, the heat rats showed that the level of secretory activity in the puberty gland was sensitive to and causally affected by changes in temperature: heat altered the cellular structure of the gonad, increasing the number of interstitial cells. In the context of the problem, the animal experiments did two things. First, they provided experimental evidence that the interstitial cells mediated cross-generational adaptations honed by environmental change; they showed that the interstitial cells were indeed the mechanism of somatic induction. Second, the flexibility in the sexual attributes that resulted enabled the authors to claim the poverty of a rigidly genetic concept of race when applied to human sexuality. The results challenged a growing anthropological emphasis on racial fixity in sex differences and argued instead for climate.

During the experiments, animals were raised and maintained in chambers in which temperatures ranged from +5° to +40° centigrade (41° Fahrenheit to 104° Fahrenheit) for up to twelve months. Control animals were raised under the ambient temperatures typical of the season in Vienna. The authors stressed their tightly controlled rearing conditions: all other influences on the groups were made equal, including light, humidity, feeding conditions, water access, animal size, and housing. Cages even had self-registering temperature and psychological gauges (*Psychographen*) that operated continuously. The results showed exaggerated sexual development in males and females reared in high but not low heat. Males had huge testes and oversized subsidiary sex organs—structures such as the seminal vesicles and prostate gland. "At the age of three months these organs had reached the volume seen in normal adult males."[11] The uteruses of very young females were mature and sometimes showed the beginning stages of pregnancy. Males and females also had early-developing sex drives. Females were in estrous and sexually receptive by eight to ten weeks of age, and young animals frequently copulated even when only about seventy days old. Over years of research, Steinach and Kammerer wrote, Steinach had never seen rats mate before one hundred days of age, and even that was early. Beyond early mating, the intensity of their sexual behavior increased throughout the animals' mature lives. Strangely, however, these signs of exaggerated sexuality in internal organs and behavior were associated with *decreases* in outer physical sex differences. External differences, like the body weights of males and females, were less in heat rats (at times five times greater in controls). Males and females showed less difference in fur density. On x-ray, their skeletons were more similar in size. It seemed that truly sexual features were exaggerated by heat, but these differences were not apparent when one observed external physical signs. Despite the rats' oversexed behavior, heat had diminished the external attributes often used to distinguish males from females.[12]

Steinach and Kammerer then examined the cellular composition of the animals' gonads. Except at the very highest temperatures, the generative cells—sperm and ova—seemed normal. But in both sexes high heat produced a huge proliferation of interstitial cells. In males, the spaces between the sperm canals were almost completely packed with "compact masses of Leydig [interstitial] cells."[13] But the effect was variable, and absolute size of the testes had also increased. So to be sure they hadn't simply chosen extreme cases, and to distinguish absolute from relative changes in the two cell types, the authors measured the thickness of the interstitial cells in up to thirty different cellular fields. In males, these were sections taken from the horizontal and vertical meridians of each testis. The authors compared these numbers to cell numbers in other areas of the gland that had also greatly increased in size. This could show that the proliferation was specific to the cell type and not due to an overall increase in gonad size. Compared to animals raised in normal temperatures, the heat rats showed a huge increase in interstitial cells, and the range of differences (minimum versus maximum) was much greater and unusually exaggerated only in some areas. If overall size changes in testes were responsible, differences should be proportionate everywhere. They were not. So the findings strengthened their conclusion that the differences were due to a disproportionate increase in interstitial cell number and not to overall increases in testis size (a criticism that had been raised against Steinach's earlier demonstrations). The results were the same in females. Steinach believed that, in females, the cells of the corpus luteum constituted the female puberty gland, and larger numbers of "lutein cells" occurred in the ovaries of females exposed to heat. In the uterus and ovaries high temperature had produced a comparable effect of proliferation in female interstitial cells.

The "Anthropological Part"

After reporting the heat rat data, the authors turned to the "anthropological part" of the paper. This was an additional forty pages in which they painstakingly developed the implications of the heat rat experiments for anthropologists' descriptions of the sexual attributes of humans in the tropics, Europe, and the Arctic. The heat rat and anthropological parts could have been two completely different papers. But in Kammerer's mind they were tightly connected; Steinach and Kammerer made it clear that they were pursuing the "homological influence" by which the endocrine mechanisms revealed in Steinach's research could help show that climate, especially tropical heat, could account for differences in the sexual attributes of Naturvölker and Kulturvölker.

The authors collected no observations of their own on people; instead they relied on the existing anthropological literature. They aimed to explain and interpret the anthropological data by what they considered a revolutionary methodological step: they would for the first time apply experimental animal work obtained under controlled conditions to settle the tangled disputes that plagued the anthropologists' observations of human beings. Throughout, the authors highlighted the many contradictory findings in the anthropological literature, especially stressing disagreements between the older and more recent anthropological observations on the impact of climate. The findings just were not very good. The measures were poor, and the data were full of confounding influences. Comparisons among peoples could not take into account differences in things like latitude, humidity, sea level, time of year, nutrition, and the types of dwellings and patterns of activity typical of people in various cultures. With so many influences at work, observational methods could never disentangle the confounds and show causal relationships. Animal experiments, however, offered definitive conclusions; and the authors presented their work as a methodological breakthrough for the human sciences, one in which controlled experiments were applied to anthropology on a large scale for the first time. The authors used "the most exacting methods of which biology is capable. . . . The more exact the application of the analytical method with animals, the sooner we may hope . . . to understand distant problems, which, by their nature, typically cannot permit comparable degrees of precision."[14] Their superior method would dispel the confusion and demonstrate the power that tropical heat had on the onset, duration, and degree of sexual maturity and sexual appearance (*Mannbarkeitserscheinungen*).[15] Not incidentally, it would also establish the developmental power of the interstitial cells reacting flexibly to environmental input.

Steinach and Kammerer rarely disputed the anthropologists' findings. Rather, their goal was to examine what kinds of processes had produced them. To do this, they described several differences distinguishing men and women among peoples inhabiting tropical versus temperate (and some Arctic) regions. Most compared Europeans to Polynesians and Africans. Following the organizational scheme laid out in their heat rat section, they examined extragenital physical differences between males and females, including growth, the distribution of hair, and the development of facial and bodily demeanor. They then turned to "subsidiary" genital differences, such as those in scrotum and mammary glands (on which there were very few observations). But the most contentious issues focused on three "reproductive" phenomena: differences among Europeans and indigenous peoples in the onset of menstruation and puberty, the degree of fertility, and the character of the "racial" sex drive. In each case, the authors argued that population differences consistently followed the rat pattern: extragenital differences were reduced, while truly genital and mental differences in sexuality were exaggerated in tropical climates.

In explaining each, they returned repeatedly to two scientific debates. The first, of course, was the relevance for humans of Steinach's puberty gland as reflected in the heat rat data. For example, summarizing measurements on Matupi Islanders, northern Europeans, and Japanese, they noted that sex differences in size between men and women were much less in warmer climates. Sex differences in adult Matupis' body lengths corresponded to those seen in twelve-year-old European children.[16] They suggested that such growth differences resulted from the inverse relationship between activity in the puberty gland and the pituitary gland: the pituitary's regulation of growth was inhibited by the climate-induced secretory work of the puberty gland.

The second theme reveals Kammerer's motivation. Following his lead, the authors repeatedly challenged younger anthropologists' racial interpretations of the differences among peoples and replaced them with the impact of climate acting through alterations in the secretions of the puberty gland: an exact parallel with the heat rats. They rejected the conclusions of Otto Reche, a German anthropologist then working in Vienna, who interpreted the small adult size and early puberty of the Matupi Islanders as a fixed racial attribute, and replaced it with an emphasis on environmental control.[17] The geographic distribution and ancestry of the Matupis made it plausible that the differences were formed by the impact of the tropical climate on development. "It is open to question whether we attribute [the effects] to climate or to race."[18] Throughout, Steinach and Kammerer consistently rejected degenerationist and reversionist interpretations of the characteristics of Naturvölker in favor of developmental accounts based in environmental difference. Speaking of the sagging breasts of African women, for example, they rejected explanations based on regression to an earlier evolutionary stage and stressed instead exaggerated development (*hochgradig hypertrophische Bildung*) that functioned to sustain suckling during continual child care.[19] Breast structure was a functionally induced adaptation, not an evolutionary reversion. They even rejected as too shallow various degenerationist accounts of Europeans that were based on the presumed damage produced by modern civilization.[20]

Race and Climate

The full force of the rejection of race and degeneration was brought to bear on their discussion of two issues: differences among peoples in the timing of their entrance into puberty; and the nature and strength of their sex drives—both, the authors believed, could be interpreted as the result of changes in the interstitial cells reacting to heat. Addressing the onset of puberty, they again noted the contradictory findings in the anthropological literature, and they again directed their strongest criticisms at younger anthropologists (like Reche) who ignored the earlier literature on the power of climate. Reanalyzing the older data to include only those observations that used exact estimates of age, Steinach and Kammerer rejected Reche's racial stance. Their reanalysis confirmed the older view: there were large differences in the onset of menstruation in regions that differed greatly in temperature. They wrote: "In fact it will prove to be most acceptable that outer and inner factors, climate (and presumably nutrition and activity) and race, condition just as well as constitution, work to influence the onset of puberty. The apparent contradictions arise from such double (endogenous and exogenous) determination of the onset of sexual maturity, which it clearly can resolve."[21]

Even more troubling were the problems of sex drive and fertility. It was widely believed that human sex drives were much stronger in warmer than in moderate or colder climates. In southern lands sex drives sometimes seemed to be uncontrollable, reflecting a complete loss of sexual morality. Even the sexual fate of Europeans in the tropics was disputed in accounts of the "acclimatization question," which explored whether Europeans became abnormally "primitive" under tropical conditions. Europeans had long perceived indigenous peoples as less civilized, inferior in mental capacity, and oversexed. But even European colonial administrators had difficulty maintaining their reserve and health in the tropics. A pervasive bodily weakness often overcame European men in African climes, as did a condition termed "tropical nervousness (*Tropennervosität*)," in which symptoms of psychic excitability or overstimulation could rise to a condition of madness (*Tropenkoller*).

The abnormal acclimatization patterns of Europeans raised vexing questions that pitted European sexual propriety against a supposed racial hierarchy of sexual control versus sexual excess. Their resolution was further complicated because the effects of tropical heat seemed to differ drastically in European men versus European women. European men in the tropics experienced a loss of control and an increase in sexual desire. They could function, though not without a loss of good old European reserve. European women had much more trouble. In the women, life in the tropics often produced severe reproductive weaknesses. Amenorrhea, uterine diseases, milk loss, natural abortions, and sterility all plagued white women who had moved to various seats of European colonial expansion. Worse, the changes could be amplified across generations, so that by the fourth generation in the tropics, European families might be rendered nearly sterile. Why were there such differences in the acclimatization of European men and women? Why did men become oversexed, while women fizzled out? What about the tropics could exaggerate or harm European sexuality?[22]

Steinach and Kammerer used double gland logic to solve these problems. In humans as in other mammals, reproduction and sexuality were not a unitary phenomenon. Anthropologists had failed to take account of the two separate functions of the double gonad: the reproductive versus the sexual. Animal evidence had shown an inverse relationship between proliferation of the interstitial cells and growth of the generative cells, suggesting that one part could inhibit the other. And the heat rats showed that heat influenced the two parts differently. Distinguishing the two functions would be especially important in understanding sexuality in the tropics.

Focusing just on heat's impact on the puberty gland proper—the interstitial cells—could account for both the supposed hypersexuality of indigenous peoples and for the odd differences between European men and European women. The first was a simple result of the heat-induced proliferation of interstitial cells on the human sex drive. If heat acted flexibly to alter human interstitial cells, "race" was secondary, and indigenous tropical peoples, like European men, were simply exhibiting the effects of tropical heat on interstitial cell number. The second depended on confusion in the kinds of measures used by the anthropologists. Unaware of double-gland physiology, anthropologists had mistakenly fused indices of the gland's two different functions, the reproductive and the sexual. This confusion had led them to greatly overstate the differences between men and women's reaction to heat. In European men, sexual drive and interest (the most common index of the anthropologists) reflected the heat-induced increase in interstitial cells; while in European women, reproductive problems like decreased fertility reflected the decreased viability of the generative portion of the gland. Both resulted from the intense tropical heat, but the first was a sexual reaction, the second a reproductive one. If, as Steinach's research showed, the relationship between the hormone producing interstitial cells and the egg and sperm "secreting" generative cells was inverse, a heat-induced increase in the interstitial portion of the gland would decrease fertility in women by diminishing the egg production (as happened in the heat rats). Animal studies had also shown physiological differences between the testes and ovaries in how gamete production was affected by heat. Maximum interstitial cell production occurred at temperatures that were about ten degrees lower in female rats than in males. So the same hot temperature might leave sperm unaffected as it increased interstitial cells, though eggs were destroyed. Both the reproductive maladies of European women and the loss of sexual control in European men could be traced to the impact of heat on the inverse relation between the two cell types. The authors wrote: "When the copulatory ability of men in the tropics increases, the child-bearing ability of women in the same location decreases. Thus this apparently divergent double phenomenon probably has a common cause: Increase in the volume of the puberty gland as a consequence of elevated temperature."[23]

Because anthropologists had measured women's sexuality using birth, fertility, and the onset of menstruation, they had wrongly equated a woman's sexual interest with her reproductive capacity. This led them to ignore a number of important implications of climate. For example, under the heat-induced conditions of infertility, a woman's interstitial cells would, like a man's, increase. The implication was, of course, that European women's interest in intercourse—their "mating ability (*Begattungsfähigkeit*)"—would increase just as their childbearing ability (*Gebärfähigkeit*) decreased. Because the neural changes in sexual motivation—erotization—would occur despite reproductive compromise, like their hypersexualized tropical sisters, reproductively challenged European women were likely to experience intense sexual desires. In both sexes, sex drives waxed and waned as a result of increases or decreases in the number of interstitial cells. As higher levels of secretion reached the brain, "erotization" caused a heat-induced hypersexualization in women, too, potentially eliminating racial differences. Moreover, in women as in men, the reproductive changes in gamete production associated with puberty occurred after the awakening of the sex drive. So the heat-induced proliferation of interstitial cell secretions could increase sex drives in women before and even at the expense of fertility and reproduction.[24] In clarifying the anthropologists' confusion, Steinach and Kammerer used the gonad's responsiveness to heat to challenge both racial differences in sexual intensity and sex differences in sexual interest. In women as in men, sexual motivation was biologically divorced from reproduction, and acclimatization actually had occurred in European men and women. Rather than inducing illness, however, Europeans' reactions in the tropics simply reflected how their gonads had adapted to tropical heat. Altered hormone secretion produced by more interstitial cells had changed the Europeans' sexual expression and erased their supposed racially based, "civilized" sexual reserve.

Steinach and Kammerer presented these conclusions in a context in which human progress and continued upward improvement were thought to depend on the differential fertility rates of peoples deemed more or less racially advanced. In this atmosphere, their double-gland logic was clearly a political hot potato. It suggested that the problems of fertility and sexuality could not be understood exclusively in terms of fixed and hierarchical racial differences. Instead, anatomically and psychologically, the sex drive and sexual expression must be separated from reproduction, as anthropologists understood sexual differences among peoples and between the sexes. Sexual restraint, acclimatization, the female sex drive, and the physiological forces driving human desire—all were produced by a mechanism that was distinct from the one governing reproduction and fertility. The author's application of the double gonad to humans even had implications for debates about access to contraceptives. The stakes that potentially turned on the double-gonad interpretation of human sexuality were huge. If, in humans as in rats, the puberty gland could be transformed by heat, almost every social hierarchy on which prevailing cultural differences depended would have to be revised in the face of Steinach's and Kammerer's interpretation.

The Heat Rats and Somatic Induction: Kammerer's Strategy

To seal their challenge to rigid racial explanations, Steinach and Kammerer always returned to the heat rats, reiterating the results of their animal experiments. Underlying them all was the now familiar (to the reader) impact of high heat on the expansion of the interstitial cells and, via the hormones produced, the early sexual development and exaggerated sex drives of the heat rats. For mammals—including humans—they wrote: "in their totality these interstitial cells constitute a 'puberty gland,' whose inner secretions bring about the maturation and maintenance of bodily as well as mental sexual characters. Therefore, the tropically enhanced variability of the gland's magnitude (*Drüsenquantität*) must entail a related changeability in the intensity of puberty, and with it a comparably increased change in its onset."[25]

But they had reported another finding in the heat rat section: one they made no special note of when they first presented the data. Early in the paper and with very little fanfare, they described interstitial cell numbers in the third generation descendents of the originally exposed heat rats. The descendents had not been exposed to unusual heat; but some nonetheless inherited glands with a huge proportion of interstitial cells. The physiological change first acquired in parents exposed to high heat had carried over to the glands of their unexposed offspring; it was inherited. Oddly, for Kammerer, and with an almost calculated caution, the authors chose to refer to this effect *not* as a hereditary one, because, they wrote, the experiments did not permit conclusions about "true heredity (*echte Heredität*)." Instead, stressing description, they stated that one could claim merely the heritability (*Erblichkeit*) of increased interstitial cell mass, which then exaggerated sexuality. But when they later questioned the hereditary basis for fixed racial differences in human sexuality, this result assumed more and more importance.

Europeans who immigrated to the tropics regularly intermarried with the natives. Sometimes European sexual characteristics persisted in mixed-race descendents; sometimes they did not. Steinach and Kammerer argued that anthropological observations on intermarriage, too, had ignored the relevance of the double gland and the cross-generational impact of tropical heat in altering it. Evidence on the persistent effects of heat in descendents of the original heat rat generation helped resolve the contradictions seen when anthropologists compared the sexual practices of Europeans of purely European heritage with those of mixed heritage who were born and raised in tropical climates. The authors examined anthropological research comparing the onset of puberty in children of purely European descent with those of mixed racial heritage, both living in tropical India. There were no differences. What had happened to the power of race? Two presumably different "races" showed the same pattern; in tropical India both entered puberty earlier than would be the case in Europe. Were these acquired environmental effects that had been inherited by European children?

Again Steinach and Kammerer appealed to the power of their experiments. Based on mammalian homology, and combined with the "purity of a laboratory experiment," the anthropological observations suggested that, once climatically acquired, the modified onset of puberty had been inherited by the descendents of Europeans born into tropical climates. Resolution of the contradictions required stepping back from the exclusively internalist perspective that relied on race to interpret human differences. Rather than race and heredity, which they termed "inner hereditary conditions (*innere, keimplasmatische Bedingungen*)," Steinach and Kammerer again stressed sensitivity to environmental circumstance that, as in the heat rats, could be manifest in descendents. When the facts were arranged in this way, they said, evidence opposed interpretation in terms of the rigid peculiarities of "race." But accepting this explanatory step depended on acceding to the idea that "climate and in general environmental factors can activate persistent organic states (*organische Dauerzustände*)."[26] Their heat rat experiments, they said, like many other experiments—presumably Kammerer's—had shown this to be possible. A view based on fixed racial differences only appeared contradictory to a climate-based approach if one disregarded the generational persistence made clear in these experiments.

In a way that appears completely disingenuous, at least on Kammerer's part, they first attempted to distance their demonstration from disputes about the inheritance of acquired characteristics, suggesting that, while the rat findings were clear, an opponent need *not* accept them as evidence of the inheritance of acquired characteristics. They wanted to *avoid*, they said, the word "inheritance (*Vererbung*)" because it unnecessarily embroiled the issue in theoretical debates. Anyone who disputed the concept of use inheritance, they said, could still accept their demonstration and view the persistent effects of temperature as mere apparent inheritance (*Scheinvererbung*), not a true hereditary change. Either interpretation, they suggested, would sustain the all-important conclusion.

But the authors quickly reversed themselves and presented the key reason why one should not do this. Their experimental demonstration, they said, differed in one important way from many others that had examined the inherited impact of environmental influence on subsequent generations. Prior demonstrations left open the question of whether the results of inherited environmental effects were due to parallel or to somatic induction. Their experiments, however, showed not just that bodily systems had changed across generations; they also showed that the changes occurred as a result of altered chemical action in a specific glandular tissue, one that was cytogenetically in intimate connection with the germ-plasm: the interstitial cells. The heat rats had provided evidence of mechanism; changes in the interstitial cells were directly proportional to the intensity of heat, they persisted in subsequent generations, and they occurred in cells known to influence the carriers of heredity. The evolutionary connection between the two cell types was, they said, deeper than that of two neighboring cells that simply shared the same gland. Rather, the interstitial cells *were* converted germ cells. Citing Nußbaum, Harms, and others, they wrote: "the first, embryological Anlage of the interstitium derives directly from primordial sex cells (*Urgeschlechtszellen*)."[27] The authors were dealing with tissues for which the separation of soma and germ had once been irrelevant. The impact of heat on the interstitial cells surely reflected a somatogenic effect. The changes they produced were somatic changes, and in mammals, the cells were clearly no longer germ cells. But the evolutionary link between the two cell types explained how heat could reach the germ via a now specialized mechanism that also transformed the soma—the hormones of the interstitial cells. "The debate about the direct or somatic induction of new attributes loses its grounding, because the debated areas partially coincide (*zur Deckung gelangen*)."[28]

Kammerer was reiterating his 1919 proposal that the interstitial cells were the physiological mechanism mediating somatic induction. But now he had the needed evidence: experimentally induced changes in sexuality produced by temperature; cellular changes in the organ responsible; and the persistence of the changed physiological system in offspring not directly exposed to heat. Kammerer's rhetoric is, however, different here. It involves a subtle caution that enables him to have it both ways. He seems to be saying: "Take the empirical result, and reject the inheritance of acquired characteristics, if you must. But be warned that if you do, given that the experiments support a physiological mechanism, their deeper impact will be lost to you." Kammerer wrote the anthropological part of "Climate and Puberty,"[29] and he must have done so with some concessions to controversial issues that Steinach wished to avoid. Perhaps in response, Kammerer deemphasized theoretical arguments about heredity, and stressed instead methodological advances in the precise use of animal experiments that could, by homology, be extended to humans. Whatever you called it, he told himself and his readers, the mechanism of somatic induction was apparent. Environmentally induced changes in a specific somatic system had been inherited; increased interstitial cells in the heat rats' descendents revealed the mechanism, and the changes it produced could be used to support the older generation of anthropologists. Race was flexible.

In the middle and late nineteenth century, a generation of German-speaking anthropologists that matured before the turn of the century had distanced themselves from evolutionary thinking, sometimes explicitly rejecting the view that indigenous peoples had no culture.[30] Indigenous peoples were more than just Naturvölker, and despite the growing prevalence of evolutionary hierarchies among peoples, this sometimes entailed the rejection of a rigid notion of race. Though he does not cite him, Kammerer would probably have known of the views of Felix von Luschan, his fellow Viennese, a pioneer in anthropological fieldwork in the German colonies, and later director of the sections for Africa and the South Pacific at the Berlin Museum of Ethnography. Stressing the historical development of both, Luschan criticized the boundaries between Naturvölker and Kulturvölker and challenged the value of race. Rejecting the exclusive study of anatomical characters, he wrote: "The question of the number of human 'races' moves entirely to the background; even the word 'race' itself has increasingly lost its meaning and would in the best case be entirely given up."[31] Luschan's student was Otto Reche. Reche held a different view, one that grew even more stridently racial when Reche espoused increasing disparities in the mental abilities of the white, brown, and black "races," indexed through physical attributes that he assumed were rigidly determined by genes.

Kammerer's rejection of race was a rejection of this increasing reliance on exclusively inner causes of heredity. While he did not entirely avoid hierarchy, Kammerer's reliance on endocrinology could be used to reject biological anthropology's new emphasis on rigid race science.[32] And some anthropologists were listening. According to Benoit Massin, anthropologists who rejected the race concept after 1910 often did so by siding with neo-Lamarckians in biology. Some even cited Kammerer's zoological research as an experimental vindication of Lamarckian inheritance. Massin quotes Fritz Hertz, who concluded that Weismannism "had received its 'finishing stroke' thanks to the 'modern' experiments of neo-Lamarckian researchers like Kammerer and Semon."[33] Kammerer's goal was to generate evidence with an even more powerful scientific foundation. His collaboration with Steinach would help restore the older anthropological view of racial flexibility by anchoring the problem of somatic induction in the secretory function of the interstitial cells. The combination made race as much a matter of the environment as of the germ-plasm.[34]

Reception of the Heat Rat Evidence

Paul Kammerer presented his hypothesis and the evidence supporting it in a scientifically persuasive way well before genetics had ceased consideration of environmental influences on heredity.[35] He framed it with the help of a well-known endocrinologist, presented it in mainstream scientific outlets, and in a way that had been suggested by many—as physiology. Prague endocrinologist Artur Biedl, a supporter of the separate secretory gland, had even referenced Kammerer's view of sexual development in the 1916 edition of his widely used textbook.[36] But in the 1920s, one looks in vain for scientific responses to the hypothesis. Examination of textbooks, empirical research on the interstitial cells, and reviews of reproductive endocrinology indicate that, even among supporters of the double-gonad concept, discussion of the interstitial cell hypothesis of somatic induction was nearly absent. Neither endocrinologists nor geneticists commented on the hypothesis one way or the other.

Kammerer was already a controversial man, so it is possible that many ignored the hypothesis because it was one of his many speculations. For several reasons, I think this unlikely to be the sole basis for the lack. One is that when the data appeared, Kammerer was still respected by many and there was interest in his work on both sides of the Atlantic. Second, he attached his idea to Steinach's name and work at a time when Steinach's star was rising. After 1920, though he remained controversial, Steinach was an acknowledged leader of reproductive endocrinology in Central Europe. The first of his seven Nobel Prize nominations occurred in 1921, just a year after "Climate and Puberty" appeared. I suggest instead that social issues, linked to Steinach as much as to Kammerer, were important in overwhelming scientific attention to Kammerer's hypothesis. Considerable controversy surrounded the scientific evidence on what, in the early twenties, was called "the interstitial cell question." Some, though by no means all, of that controversy was social. Embedded in the scientific debates were the implications of the interstitial cells for a biological understanding of "the women question" and for the nature and extent of biological differences between men and women. These too, affected the reception of the hypothesis.

The Interstitial Cell Question

Several debates about the role of the interstitial cells turned on genuine scientific differences. Disputes about evidence were linked to problems produced by differences in, for example, the use of proper tissue staining techniques, variations in the timing of development, and species differences in the animal tissues used. These debates centered on three interrelated questions: were the interstitial cells secretory cells or merely passive connective tissue that might nourish or protect sperm; were the cells derived embryologically from the developing cell layer termed the germinal epithelium (which also produced gametes), or did they originate in an unrelated layer of tissue; and finally, did transplant or vasectomy actually increase the number of interstitial cells, as Steinach believed, or was the effect an artifact of altered gonad size or other influences?[37] Those who rejected the secretory function of the cells usually took the gametes to be the primary source of the secretions. This made it possible to interpret Steinach's findings either in terms of gametes that were still active in altered or transplanted gonads, or in terms of substances produced when the gametes died and their products were reabsorbed.

But far more than science was involved. As objects of scientific analysis, the interstitial cells also embodied profound social questions. Perhaps the most inflammatory was the issue of how biologically separate and specialized the sexes were. Gametes had traditionally defined the differences between the sexes: as we saw in Chapter 2, the passive female egg and the active male sperm were reified into broad views of masculinity and femininity. If the sex hormones were produced by those gametes, the thinking went, the hormones' effects would be as separate and distinct as the gametes. Sexuality would continue to be defined by the demands of reproduction, and different gametes still could be used to justify sharp cultural differences between men and women. But the double-gonad concept proposed the functional separation of sexuality and reproduction—of hormone and gamete. If interstitial cells produced the hormones controlling sexual development, then sexuality was not a result of differences in gametes, and it might vary in ways not reflected the gametes' reproductive functions. The passive, disinterested, nurturant female personality would, in effect, lose her biological foundation, as might the active, aggressive, decisive male. The social implication was that despite striking sex differences in reproductive specializations (gametes, pregnancy, and child birth), the sexual desires, characteristics, and motives of men and women were far more similar than previously thought. Related proposals of endocrine bisexuality and the cross-sexual impact of the hormones (see Chapter 2) just aggravated the problem.

Scientists fell out on all sides. But it is telling that even proponents of Steinach's view that the interstitial cells secreted sex hormones failed to mention Kammerer's hypothesis. Some of the latter even agreed with Kammerer's view of an intimate connection between the gonad's two cell types. In his 1924 review of reproductive endocrinology, for example, Dorpat and Chilean endocrinologist Alexander Lipschütz stated: "The interstitial cells are elements which are confined to the generative glands, and they are no less specific than the epithelial generative cells themselves. How is this to be explained? The most likely explanation seems to lie in the suggestion that they are autochthonic elements formed on the spot and originating from the same primitive tissue as the generative cells."[38] Lipschütz noted the connection, but he did not link it to somatic induction. In 1923, Yoshitaka Kitahara published a comprehensive analysis of the interstitial cell question that added measurements of cellular changes in human gonads occurring during prenatal development. Kitahara worked in Walter Kolmer's laboratory in the Physiological Institute at the University of Vienna. Kolmer, too, studied the interstitial cells, and given the lab tradition set by Exner, it is reasonable that the Institute would be sympathetic to Steinach's approach.[39] And Kitahara was. He cited Steinach repeatedly as he discussed the numerous problems left unaddressed by critics of the secretory function of the cells. Identifying "the Viennese school" as the major proponent of the cells' secretory function, Kitahara concluded that the interstitial cells served both a protective and a secretory function, and they could influence sexuality even when the gametes had atrophied.[40] He even stressed the reciprocal influences of the two cell types. Without the influence of the germinal tissue during development, the transformation of the interstitial cells into functioning gonadal tissue did not occur. And, once developed, the interstitial cells became mediators (*Vermittler*) between the gametes and the soma. In keeping with Kammerer's hypothesis, therefore, Kitahara acknowledged a dynamic formative relationship between the two cell types and a secretory function for the interstitial cells. But he did not reference

Steinach and Kammerer's heat rats, nor did he mention the proposed connection between the interstitial cells and somatic induction.

Neither Lipschütz nor Kitahara was interested in the inheritance of acquired characteristics; they addressed histology and endocrine function, not heredity. So the omission might reflect their explanatory focus on endocrinology. But one of the most vocal opponents of the secretory function of the cells was also a man deeply interested in heredity and in the inheritance of acquired characteristics: the German anatomist Hermann Stieve. In the early 1920s, Stieve's interests focused on the impact of environmental conditions on gametes, particularly those that affected fertility by damaging the germ cells during egg and sperm formation. Like Kammerer, Stieve was an expert in amphibians, and his emphasis on the environment drew him, too, to the problem of the inheritance of acquired characteristics (though he denied the possibility of somatic induction). Despite their common interests, however, politically and scientifically, the two men stood on opposite sides of every theoretical issue surrounding the interstitial cells. Kammerer and Stieve also had very different fates. Eventually a specialist in the anatomy of female reproductive organs, Stieve became a huge success. He was rector of the University of Halle, and in 1935, he was named director of the Anatomical Institute at the University of Berlin. There, during the Nazi period, he conducted experiments on women who had been executed in the infamous Berlin-Plötzensee prison. By the early 1940s, he was one of the most well-respected anatomists in Germany.[41]

In 1921, as a young scientist, Stieve published a massive review on the interstitial cells.[42] In it he was sharply critical of Steinach and the possibility of a secretory function for the cells, a proposal that he identified with "the Vienna school"—a label he coined. He focused his rejection on Steinach's and Tandler's findings, arguing that their physiological approach paid insufficient attention to microscopic cellular anatomy. Concentrating on his own work, he challenged the idea that the interstitial cells could be accorded "a higher meaning of central importance for the development of the organism."[43] Instead, he argued that sexual differences were fundamental to an organism and resided entirely in the generative cells. The interstitial cells functioned as supportive cells, which merely maintained healthy nutrition for sperm and eggs. And he consistently rejected Steinach's demonstration of an increase in interstitial cells in the testes following vasectomy. Without mentioning Steinach and Kammerer's argument supporting an absolute increase in interstitial cells with heat, Stieve treated the effect as an overall compensatory increase in the total number of cells in the gonad.

Several scientists who supported Steinach criticized Stieve's analysis. Knud Sand had, like Steinach, done transplants on mammals that confirmed the secretory function of the interstitial cells. He used different surgical procedures that nonetheless confirmed the effect of increased interstitial cell numbers on the development of masculine and feminine sexual features. He sometimes even criticized Steinach, rejecting, for example, Steinach's notion of a glandular antagonism between ovaries and testes. But good evidence supported the secretory function of the interstitial cells, and he seemed puzzled that there was still controversy about it, especially in Germany. Some of the opposition was "rather biased," and, he wrote, "many of these articles . . . contain astonishing misunderstandings and erroneous reports."[44] Although he admitted that important details remained unresolved, the basic concept had been more than confirmed. Lipschütz (whom Stieve included in "the Vienna school") also presented persuasive evidence that refuted many of Stieve's objections on the relative versus absolute increases in interstitial cell number. He showed that the insertion of small fragments of grafted testicles with proliferating interstitial cells was sufficient to ensure the normal sexual development of guinea pigs.[45] Overall numbers of cells did not matter; it was their secretory activity not their number that altered sexual development. But Stieve was not convinced.

Perhaps the most balanced and comprehensive position on the interstitial cell question was the review published by Jürgen Harms in 1926.[46] The title of Harms's book, *Körper und Keimzellen (*Body and germ cells*)*, does not reveal the broad thrust of its scope. The book's central theme is the relationship among gonads, sex, and various problems of sexuality. It is a testimony to how much interest there was in the relation between soma and germ almost forty years after Weismann's proposals, and to how much of that interest turned on understanding sex determination and the endocrine basis of sexuality. Harms dealt with the germ cells of the title, but primarily in relation to the chromosomal (or not) determination of sexual attributes and behavior. He also addressed the problem of the "bisexual potential (*Veranlagung*)" of animals, including sections on homosexuality and hermaphroditism; he reviewed the relation of the gonads to other glands in the endocrine system; and he summarized the evidence of external influences, such as toxins and alcohol, on gonad function.

Harms paid close attention to the interstitial cells, discussing evolutionary differences in the presence, number, and cycling of the cells. He presented evidence on both sides of the debates surrounding the cells, including their evolutionary and developmental origins and their secretory function. If the cells were of epithelial origin (as the germ cells were), then it would be reasonable to accept them as undifferentiated germ cells (*Urkeimzellen*).[47] If they functioned purely in nourishment (as Stieve and others insisted) or to mop up degenerated germ cells and their products, they should have no influence on sexuality. If, on the other hand, they were secretory cells, they could be key players in shaping many manifestations of sexuality.

Harm's treatment was far less polemical than Stieve's. He often reached no firm conclusion, leaving the reader with the statement that more evidence was needed to complete the picture. In reviewing the secretory question, for example, he noted several features of the cells' structure that were consistent with a secretory function (they had fats deposits like other glandular cells and they were often positioned near the circulatory system). Yet Harms sided with Stieve; he invested sexuality and reproduction entirely in the gametes. When presenting positive evidence for the secretory function of the cells, he would often interject that a glandular function for the cells was "doubtful" or "questionable." His descriptions of Stieve's work rejecting a secretory function showed less caution; he used words like "certain," "clearly," and "without objection." He viewed the interstitial cells as a "metabolic apparatus of the testes." They functioned in the reabsorption of hormones and their leftover products, and he was convinced that "a special puberty gland in Steinach's sense" must be rejected.[48] "In both the male and the female gonad the sex cells . . . alone serve reproduction; all other elements . . . are to be interpreted only as accessories (*Hilfszellen*) to the maintenance of the development of the gametes and the sex cycle. Through the research of Steinach and his school, particularly the interstitial cells have attained a meaning that one should certainly not entrust to them."[49]

Harms repeatedly returned to the question of the relationship between soma and germ cells. Some research had explored relationships in the developmental timing of the appearance of the two cell types. Other work showed how external circumstances or substances like alcohol and radiation affected the tissues of the gonad. These discussions are the most obvious place to find reference to Steinach and Kammerer's collaboration, and indeed there is one.[50] Paying close attention to the heat rat experiments, he the stressed findings that supported Stieve (decreased fertility with the highest temperatures)! He criticized Steinach and Kammerer's method of measuring increases in interstitial cells, and instead described Stieve's work on mice, which he considered more solid.[51] He made no mention of the issue of somatic induction. Focusing on the decline in fertility, which he (and Steinach) interpreted in terms of degenerated gametes, Harms then made the only reference I have found to Steinach and Kammerer's anthropological discussion. He was curt and dismissive: "As Steinach and Kammerer point out, all of these questions also have implications for the issue of the relation between climate and puberty in humans. On the comparison of the preceding experimental [rat] results with the ethnological and anthropological findings on human populations of warmer climates, I do not comment here, because they are not yet ripe for decision."[52]

Harms's social views are not as clear as Stieve's. And though he sided with Stieve, he also built a case for the inconclusiveness of the science on many issues. But the decisiveness with which he accepted Stieve's conclusions suggests that he too was uncomfortable with the degree of sexual malleability implied in Steinach's program. He illustrates well the impact of the interstitial cell question on Kammerer's hypothesis. Despite his attention to climate and elevated temperatures, Harms did not mention the impact of heat on the unexposed descendents of the heat rats; nor, despite Stieve's interest in the inheritance of acquired characteristics, did he mention the proposed link to somatic induction. Because the opposition did not accept a secretory function, evidence on the hereditary persistence of the modified secretory structure in relation to sexuality was unremarkable. If the cells had no secretory function, all of that was irrelevant or premature. The other problems surrounding the interstitial cell question had rendered invisible or irrelevant just that aspect of the Steinach-Kammerer collaboration that Kammerer considered most significant: the relation of the cells to somatic induction and therefore to racial malleability.

The Double Gonad and Attitudes about Race and Gender

Arguably the most important basis of scientific debate over the cells was the relative merit of physiological evidence versus evidence on the cells' microscopic anatomy and origin in resolving the issues. The two kinds of evidence came from two fields that had become distinct specialties with different methods, definitions, and assumptions. As a result, their contradictory conclusions could rarely be settled with more evidence. In such circumstances, social biases can play a great role as they fill the many gaps left by missing evidence. Hermann Stieve was convinced that a biological division of function separating sexuality and reproduction could not occur in "higher" animals. His motive seems to have been to retain an absolute barrier between male and female as categorically distinct groupings of life at all levels. Stieve, for example, also rejected Tandler's evolutionary hypothesis of asexuality. Tandler argued that sex differences evolve initially out of neutral (asexual) characters, which represent pure species attributes. For this reason, he believed that castration returned mammals to a neutral state. But Stieve objected, "there is no asexual species form; the lack of gonads is always a diseased condition."[53] In the midst of a debate about the relative merits of histological versus physiological data on the interstitial cell question, this point seems out of place. But as the "Vienna school" moved toward relative concepts of masculinity and femininity, Stieve was intent upon rejecting any notion of biological asexuality or bisexuality, evolutionary, developmental, or cellular.[54] Organisms were either masculine or feminine to their cores.

Stieve's conservative morality, his view of the role of women and its relation to the German population are reflected in a paper that he published in the mid-1920s on infertility. In it he expressed great concern that Germany might begin to experience the large population declines then apparent in France.[55] To explain the trend, he dealt with the causes of acquired childlessness, especially as they resulted from constitutional defects among people who would like to have children. He stressed one of the central themes of his animal research: the impact of adverse outer conditions that limit reproduction, especially in women. He had shown that alcohol was toxic to germ cells. But here, he also stressed less drastic effects that could reduce fertility, and might go unnoticed in overly cultured urban people. He mentioned the effects of tropical climates and high heat and the impact of confinement or fear, a foreshadowing of his later work. In a section on the impact of "Civilization (*Kulturleben*)," he stressed the problems of large cities, where, he said, lewd books and too many movies were among the reproductively destructive dimensions of urban living. Men could recover from these effects, he argued, but they were much more damaging for women's germ glands. Women were more sensitive—even thinking about sexual activity could damage the uterus. If girls were subject to such products of civilization, their entire development could be retarded. The mental effort involved in education and university study, he said, produced an especially steep decline in a woman's fertility. Urban women who took on manly jobs, Stieve said, even suffered from a disease called *Metropathie*.

Stieve closed with a nationalistic warning. The Germans, he said, have a great drive to preserve themselves. In the past, this force had increased the size of their population, and this had made it possible to withstand violent assaults. But, he warned, "If the will and the force for self maintenance is now lost, then the German people will in a few decades succumb to the greater power of those states in which this will is lastingly proclaimed through strong population increase."[56] The solution required caring for the Fatherland by producing many German children. For Stieve, doing so entailed avoiding the dangers wrought by weakening the traditional division of labor between the sexes. The integrity of women's reproductive roles guaranteed the protection of the very eggs that would become these German children. And the ovum as a source of femininity ensured the firm biological base of these roles.

By the 1940s, the idea that the interstitial cells might produce sexual secretions that could shape femininity separately from reproduction was much more widely accepted in endocrinology. But Stieve still rejected it; for him, sex hormones were produced by gametes that preserved the gender roles on which the future of the nation depended.

Hermaphrodites and Social Roles

The social problems posed by the desire to maintain strict boundaries between masculinity and femininity and their relevance to the interstitial cell question are well illustrated by a surgical procedure performed at the Women's Clinic in Innsbruck on September 2, 1921. A thirty-year-old woman sought relief from the pain brought by what was assumed to be an inguinal hernia. On physical examination, it became clear that the individual seeking relief was a hermaphrodite; despite many feminine anatomical features, she also had testes and a scrotum. The person had been raised female and, up to that point, lived as a woman. But her glands revealed her to be a man. She was treated with castration, and her case became the focus of an article in a major scientific journal. After the surgery, the author, Felix Sieglbauer, examined the state of the sperm and interstitial cells in her excised testes. His interpretation gives us a picture of the magnitude of the problem posed by Steinach's view of the interstitial cells.[57]

The testes showed signs of inhibited sperm production. Some sperm heads were present, but few had developed and many were degenerating, as were the seminiferous tubes. But Sieglbauer noted substantial increases in the interstitial cells. The cells had infiltrated the entire interstitial area, like a tumor. Citing both sides of the interstitial cell debate and noting that the issue was not settled, he agreed with those who concluded that the interstitial cells were glandular cells. He reviewed Steinach's position on the inverse relationship between interstitial and germ cells and the dependence of secondary sex characteristics on the former. But he also noted that Stieve and others held that in hermaphrodites the critical secretions were produced by germ cells. Sieglbauer was cautious; rather than drawing a firm conclusion, he noted two issues that prevented consensus. One was discipline. The histologists argued that the "secretory" role of the interstitial cells was nothing more than supportive nourishment. Physiologists, on the other hand, followed Steinach and viewed the cells as the critical sources of secretions controlling the development of masculinity and femininity. The second group, he said, could, however, not offer conclusive evidence of the ancestry (*Herkunft*) of the cells. Only with this would we have "a clear picture of the meaning of the interstitial cells." Seeming to prefer the inclusion of a stronger hereditary element in sexuality, Sieglbauer stressed Goldschmidt's intersexuality rule, which united Mendelian determinants (genes) with physiological ones and stressed the importance of developmental timing as a cause of the mix of male and female characters in hermaphrodites. This idea led Sieglbauer to conclude that both cell types might be involved.

Without any resolution, in concluding, Sieglbauer turned to the human question of what sex should now be ascribed to his patient. He noted the presurgical disconnect between the gonads present and the patient's psychological and social preference. In terms of gonadal structure, his patient had been a man. The glands were testes and their gametes were sperm. But the patient had been baptized and raised a woman, and "the error of sexual assignment had more or less asserted itself in consciousness."[58] Sieglbauer added the "strange" fact that after the operation, once emasculated, his patient felt good—better than ever before. His final statement illustrates the crux of the social problem. Noting the remarkable legal and social fate of such individuals, he reported that freed of "his" testes, the patient had (despite those pesky Mendelian determinants) jubilantly entered into marriage with a man.

The debates about the interstitial cells went well beyond scientific issues pitting histological evidence against physiological (though these were real). They were also infused with the controversial issues raised by categorical and hereditary versus continuous and flexible boundaries between masculinity and femininity illustrated by Sieglbauer's case. But they also framed concern about the causes of homosexuality, the possibility of new means of birth control, and the effectiveness (or not) of Steinach's rejuvenation procedure. All were intertwined with "the woman question" and with the implications for women of the biological separation of reproduction and sexuality. Steinach's demonstration of the sexual power of the interstitial cells freed femininity from exclusive definition in terms of reproduction, especially in the crucial domains of mental and behavioral sexuality.[59] As late as the early 1960s, writing from Chile, Lipschütz confirmed the controversy surrounding Steinach's work: "People, including scientists (the most narrow-minded caste in human history!), became frankly alarmed when the Bouin-Group and especially Steinach began talking about the interstitial gland, i.e. about there being a special gland producing sex hormones, a special part of the sex gland not coincident with that responsible of [*sic*] reproduction, and so to say as they thought, justifying the then beginning custom of birth control."[60]

Secretory interstitial cells removed these issues from a sharply boundaried view framed by the ability to make eggs and babies and placed them in a gray area, where chemicals that could "make" masculinity existed in females and those that could "make" femininity existed in males. The cells had become epistemic objects, their scientific status a field for debates about changing definitions of masculinity and femininity that pitted conservative views of sexuality against progressive ones. Kammerer's interstitial cell hypothesis of somatic induction critically depended on the secretory function of the interstitial cells and on the scientific communities' attention to the evidence from the heat rat's descendents. Scientific consensus began to form after his death.[61] In the early 1920s, however, Kammerer's idea and the evidence supporting it were lost in the socio-scientific storm surrounding the social implications of the interstitial cells. He had gone beyond the inferences warranted by scientific consensus, and the social implications of the physiological separation of reproduction and sexuality overwhelmed serious attention to his hypothesis.

5

"Productive" Eugenics

Harnessing the Energies of Development

Paul Kammerer was a visionary—so said his friend Hugo Iltis. He lived at a time when utopian visions of the new science of eugenics were emerging in biology, and his ideas on biology as a positive force in social evolution began to appear as early as 1910. In 1913, in *Sind wir Sklaven der Vergangenheit oder Werkmeister der Zukunft?* (Are we slaves of the past or masters of the future?), Kammerer expressed his enduring hope for an improved humanity.[1] The advances he envisioned went beyond better living conditions, good housing, good nutrition, health, and medical reform. For Kammerer, life was malleable at its core; external influence offered a lever for change—a way to become the craftsmen of a more just world. The changes he imagined did not just apply biology to society; they went further to leverage a productive power inherent in life and harness it to craft a biologically more ethical humanity. That power was the power of development. Natural science, Kammerer said, could fulfill the task of "unleashing the tremendous forces that are still bound in human plasma in the direction of forward development."[2] It was a eugenic vision; Kammerer advocated eugenics and shared many of its goals. But his eugenics was one that privileged development. For this reason, he downplayed selection, instead stressing methods that differed radically from those prevailing in the mainstream eugenics of the 1920s and 1930s.

Eugenics now serves as the epitome of science gone ethically awry. But in Kammerer's youth, it was accepted as the ultimate promise of scientific progress, a promise widely shared by international scientists on all sides of the political spectrum. Kammerer assimilated this promise and some of the hierarchy of the concept; but his approach was distinct from the version that flourished in America in the wake of World War I. He eventually became sharply critical of what he called "selectionist" eugenics. His vision for human improvement aimed to replace the socially imposed selective elimination of the "unfit" with a eugenics based in the inheritance of acquired characteristics and guided by the view that improvement in the human condition must respect the inherent worth of all individuals. Rather than sacrificing supposedly "unfit" individuals for the group, he saw the fates of individuals and the fates of groups as organically connected. He believed that a eugenics guided by environmental change that respected individual rights could even shape a human altruistic impulse, one already present in incipient form, and mold it into a full-blown biological drive for the general good.

Kammerer used his science to build a network of concepts into the intellectual base needed to justify and promote his alternative eugenics. These concepts included the inheritance of acquired characteristics; a stress on symbiosis, or mutual aid, as a positive force in the process of evolution; and the biological tradeoff between fecundity—sheer numbers of potential young—and the quality of resources devoted to the care of fewer offspring. Their integration was the basis of his "productive eugenics," and central to that integration was Kammerer's approach to the relationship between heredity and development. After 1910, mainstream eugenics paid scant attention to development. Productive eugenics, on the other hand, was based in Kammerer's belief that the malleability inherent in life's developmental processes included changeable heredity; through it, science could remold humanity in ways dictated by altered environmental conditions that reached to life's core. For Kammerer, the approach was more scientifically valid, more effective, and more ethical than the selectionist methods of most mainstream eugenics. In the context of the socialist values shaped by a largely Jewish intelligentsia in Vienna during the interwar years, Kammerer's commitment to social reform via development intensified his desire for an alternative eugenics that rejected the forced sterilization and fixed racial hierarchies that increasingly dictated mainstream eugenics.

Selectionist Eugenics

Eugenics was proposed in England in the 1870s by Charles Darwin's cousin, Francis Galton. Galton had been inspired by the work of his famous relative to found a new science that would apply Darwin's mechanism of selection to the selective breeding of "good" humans. Before 1900, that science had been largely theoretical. But after the turn of the century, the international eugenics movement grew in influence with the rediscovery of Gregor Mendel's laws of heredity, the development of the chromosomal theory of genetics, and increasing public concern about perceived threats to humanity. As an applied biology dedicated to the improvement of the human species, it flourished when scientists celebrated Charles Darwin's one-hundredth birthday in 1909. That year, Darwin's son Leonard endorsed a vigorous attempt to achieve Galton's promise, and the movement solidified.[3] But the Americans had already begun, and eugenics grew especially rapidly in America, where the implementation of eugenic measures begun before the war served as an international test case for the movement's potential. Europe watched as more and more American states endorsed legal compulsory sterilization of those deemed "unfit."[4]

Eugenics was linked both to Darwinism and to the new genetics. Darwin's proposed mechanism of natural selection is based in the idea that variation among the attributes of individuals provides the raw material for a process of natural selective breeding. As that process unfolds, individuals unable to compete for resources are removed from the breeding population of a species. An animal that cannot, say, find its food as efficiently as others, will have fewer of the resources needed to produce and sustain young. The more successful individuals will, then, leave more reproducing offspring. In doing so, efficient competitors inadvertently ensure that future generations are dominated by their inherited attributes. The ineffective beasts' reproduction is hampered, and fewer of their offspring are around to define the future attributes of the species. This differential reproduction produces species change.

The logic of natural selection assumed that selection could not produce new variants; it could only perpetuate (or eliminate) preexisting ones. Staunch neo-Darwinists eventually assumed that the sources of variation that fed evolution via natural selection were based in changes in the germ-plasm. In nature, on this view, it was the chance relationship between fluctuations in the resources needed to reproduce and the occurrence of variation—only randomly related to those resources—that shaped differential reproduction and, therefore, the species. "Nature" (a metaphor for struggle and competition) did the selecting from among the array of varying attributes represented by the competing individuals.

Darwin stressed the parallel between artificial and natural selection. And early on, Galton saw a need for human artificial selection. His statistical notion of regression to the mean meant that without intervention, human populations would long remain mediocre; natural selection would be slow, and the continued breeding of inferior races and less intelligent and/or degenerate individuals would drag down the human average. Viewed in this way, evolution largely involved blind luck in which nothing in nature guaranteed progress. Decline was, in fact, likely. The huge changes brought by urbanization and industrialization led social critics to fear the absence of any natural guarantee of progress. Indeed, they usually assumed the opposite; natural regression was compounded by decay produced by humane social measures. Advances such as medicine, charity, and welfare were measures that effectively perpetuated "the weak"—those marked with what was called "hereditary taint." From the point of many, society had violated nature, allowing the "weak" and "inferior" to breed more. "Thus modern charity, modern philanthropy, and modern medicine combine to interfere with that *selective death rate* which, biologists tell us, has, hitherto played an important role in race-betterment."[5] Because such humane measures were "counter-selective," society had inadvertently become an agent of decay. Along with increasingly negative views of the poor that were fed by urbanization, the rise of the middle class, and the new criminology, many scientists saw inferior "types"—"types" steeped in hereditary degeneration—as out-reproducing the paragons of humanity: white, middle-class, Western male professionals.

But if the reins of selection were placed in the hands of scientists and reformers, controlled selective breeding could produce progressive change in humanity, just as it had in plants and animals. Eugenic intervention would harness selection to shape the heredity of the human species. This was the promise of selectionist eugenics. As it flourished after World War I, "positive" and "negative" selection became the basis of mainstream eugenics. Positive eugenics was devoted to perpetuating greater reproduction of those individuals presumed to represent the "best" in human heredity; while negative eugenics was devoted to eliminating the "weak," infirm, or "unfit" from the breeding stock. Galton himself favored the former, but the eugenics that flourished in America and in Germany between 1910 and 1945 favored the latter. Both aimed to regulate human breeding using social and medical controls that ensured that the "fit" would out reproduce the "unfit."

Following the devastation of World War I, middle-class birth rates dropped in all of the major capitals of Europe, and many of the most courageous young men—sons of what were taken to be the "best" families—were killed in war. The death, destruction, and disease produced by war fueled fears that the tragedy would hasten the degeneration of humanity. Many individuals added war to a growing list of the "dysgenic" effects of modern life, and intellectuals and activists across Europe and America (many of them pacifists) urged more direct eugenic action. Most eugenic initiatives that were pursued after the war were based on two assumptions. The first was acceptance of the view that natural selection was the primary mechanism of evolution; this made selective breeding the appropriate means by which society might take charge and eliminate the threat of the "unfit." The second was a set of hereditarian designations (whose targets varied from nation to nation) of the attributes that defined who was "fit" and who was "unfit." By 1920, many eugenicists accepted a genetic determinism that defined the characters used to judge presumed inferiority or degeneracy—attributes or conditions such as insanity, feeblemindedness, pauperism, epilepsy, colorblindness, criminality, alcoholism, and depression—leaving little or no room for flexible developmental processes that might give rise to these attributes. Given these assumptions, fertility and reproduction could be regulated in a number of ways: through birth control, immigration restriction, marriage laws, forced sterilization (first enacted in the United States in 1907), or, in the Nazi example, by murder. But few of the abuses of the future were apparent in 1920, and many well-meaning scientists saw eugenics as a means to save humanity from the dangers wrought by modern life.

Degeneration versus Regeneration

Kammerer supported the goals of eugenics—human hereditary improvement at a fundamental (species) level. But his eugenics was based on the inheritance of acquired characteristics. To understand the novelty and the ethical superiority that he saw in "productive" over "selectionist" eugenics, we must return to the concept of degeneration. After the turn of the twentieth century, many physicians and biologists gave the environment a negative role in enhancing the decay of human heredity. The psychiatrist Emil Kraepelin, for example, argued that "degeneration insanity" was caused by dirty industrial settings and crowded urban environments. Further, environmentally based deterioration was not confined to the affected individual. Kraepelin made it clear that the toxic and demoralizing habits of urban life also produced progressive deterioration in the germ-plasm of the race. The danger was both acquired and inherited—part of the hereditary burden of modern life (see Chapter 2).

Such negative uses of the inheritance of acquired characteristics were common in early twentieth-century biology and medicine. They received social support through perceived threats to the inevitability of social progress, but they also had support from within biology. They were sustained by the debate about parallel versus somatic induction and by the widely accepted theory that ontogeny recapitulates phylogeny. Most negative environmental influences, things like alcohol, toxins, and radiation, were thought to act through parallel induction; that is, they directly altered germ and body in parallel, and their impact damaged many organs at once. In theory, nothing prevented specific cells or organs from adjusting to positive environmental change or from compensating for damage in ways that carried over to the germ. But many experts who accepted negative influences, understood as parallel induction, denied that adaptive, that is positive, adjustments could be inherited through somatic induction. In effect, they denied that the environment could have a positive impact. The result was an asymmetry in the perceived role of the environment that privileged its potential for decay and ignored or denied the possibility of environmentally mediated positive change.

The idea that ontogeny recapitulates phylogeny, a view then called the fundamental biogenetic law, had much the same effect. The law held that the stages of development seen in an individual's lifetime (ontogeny) were a rehashing or recapitulation of the evolutionary history of the species (phylogeny). Though most saw the process as offset by functionally sustained species changes, if recapitulation were true (Kammerer and many others accepted some version of it), any evolutionary progression was built on the possibility of reversion to older—less evolved—ancestral steps or stages. As a result, reversion to an older, "inferior" form could be a corollary of the process of evolution. And negative environmental influences, like alcohol use, could effect a change in germ cells that triggered a hereditary backslide to the beastlike attributes that lurked in the history of evolution.[6] Coupled with the assumption that degeneration resulted from the disuse of vital organs and the fear that positive change introduced teleology (as discussed in Chapter 3), these ideas linked neo-Lamarckian inheritance to degeneration. In principle, of course, evolution would also have positive consequences. But after the turn of the century, eugenicists rarely considered the positive impact of the environment. Rather, they saw the environment as triggering evolutionary degeneration via reverse recapitulation, and most believed that selection among fixed heredity "particles" was the best means to offset that to ensure human betterment.

The threat of negative environmental effects was intensified by the growing influence of Weismannism. The "tainted" pedigree of the American family pseudonymously called "the Jukes" offers a striking example. In 1877, American social reformer Richard Dugdale described the inherited defects present in generations of the descendents of an upstate New York family that was steeped in poverty and overridden with prostitutes, drunks, and "misfits." The Jukes were presented as a case history of hereditary degeneration. But in 1877, heredity and environment were not yet so separate, and Dugdale expressed great hope that placing the family in a healthy environment would reverse the hereditary trend. Dugdale died suddenly in 1883. In 1916, as U.S. eugenics thrived, the Jukes were revisited. Arthur Estabrook traced more than two thousand of the Jukes's descendents—including many new ones. He reported that the family was in worse shape than ever. Misfits abounded, and any hint that negative attributes might be influenced by environmental reform was gone. In the preface to Estabrook's study, *The Jukes in 1915,* the leading American eugenicist Charles Davenport wrote, "The special opportunity that the present investigation afforded was to note the later history of these strains in the presence of great changes of environment." That opportunity showed that because the Jukeses usually married "like themselves," changing environments had had no impact. Davenport even warned that scattering the family members geographically and exposing them to "better stocks" would allow their genetic weaknesses to contaminate "better germ-plasm." It was "like scattering firebrands—each tends to start a fire in a new place."[7] By 1915, for many, hereditary processes had become impervious to environmental input—they had acquired a distinctly Weismannist base.

Well after 1915, however, Kammerer consistently stressed the productive and even therapeutic power of the inheritance of acquired characteristics. Indeed, he used the term "adaptation" to refer to an immediate positive accommodation by an individual to an environmental change, no matter how it was produced or how long it persisted. Pointing out the asymmetry brought by the pessimism that environmental effects could only add burdens, Kammerer distinguished between the positive and the negative effects of the inheritance of acquired characteristics. He compared the hereditary burden (*erbliche Belastung*) produced by negative environments to the removal of hereditary burden (*erbliche Entlastung*) and to the production of novel hereditary assets made possible, not by selection but by the inheritance of acquired characteristics.[8]

This emphasis is partly the result of the Vivarium's foundation in Wilhelm Roux's developmental mechanics (see Chapter 3). Vivarium director Przibram had stressed the connection between the laboratory and exploring external influences on development in his announcement of the lab's founding in 1902, and much Vivarium research focused, not on degeneration but on biological *regeneration*, using techniques such as organ removal and transplant in young organisms. Such modifications redirected the "energies of development" to produce adaptive novelties.[9] As a product of this research milieu, Kammerer naturally stressed symmetry in the hereditary power of development. And his research consistently championed the positive (adaptive) power of the inheritance of acquired characteristics.

Kammerer's Productive Eugenics

What sometimes appear to be unrelated and disconnected aspects of Kammerer's career converge as comprehensible if each is seen as part of the logic of how novel adaptive attributes, and therefore true progress, might evolve. Kammerer's extension of biology to humanity depended on three seemingly unrelated concepts. First is his passionate commitment to the inheritance of acquired characteristics (Chapter 3); second is his advocacy of symbiosis (*gegenseitige Hilfe*) as an equal partner to competition in producing evolutionary change; and third is his analysis of the inverse relationship between the quality and the number of offspring produced by a species. Each operated in nature, and with them he built a case that no true progress would be possible if evolution were based only on competition and natural (or artificial) selection. Instead, to craft a "productive eugenics"—one that produced novelty—he privileged the modification of development and developmental flexibility, dimensions of life that were all but ignored by mainstream eugenics.[10]

Kammerer built the logic of productive eugenics most completely in two works: *The Inheritance of Acquired Characteristics* and *Neuvererbung oder Vererbung erworbener Eigenschaften: Erbliche Belastung und erbliche Entlastung*.[11] In what he termed the "eugenical part" of the English work, Kammerer compared the power of productive eugenics to what he saw as the poverty of selection. First, he showed the insufficiency of selection: selectionist eugenics was not productive; it was *re*productive. It could repeat or multiply an attribute that was already present in a population, but it could not produce novelties that transcended the throw of the dice. Citing his countryman Geza von Hoffmann's *Race Hygiene in the United States,* Kammerer contrasted the negative selectionist eugenics put into practice in the United States with the methods, assumptions, and power of productive eugenics. He asked of selectionist procedures such as deportation and forced sterilization: "Did not selection . . . succeed in creating something new?"[12] His answer was a clear "no, it did not." The changes produced might affect the composition of a population, reproducing more of some attributes and fewer of others, but they could not modify the range of available attributes. "Nothing essentially new would be created. . . . Selection falls down when required to accomplish something bigger than that which already exists."[13] Kammerer did not deny that selection could act or that negative environments could have negative impacts; rather, he argued that these outcomes could be altered by the power of development supported by an improved environment, one that produced novel inherited attributes. Truly novel and progressive change could never result from selection. That was possible only through developmental control, which induced novelty through the inheritance of acquired characteristics. Exercise, nutrition, glandular modification, climate, education—all could produce adaptations not available to selection once the changes were inherited by the descendents of those in whom new attributes were first acquired. Virtually all of Kammerer's animal research confirmed this; and he saw no reason why in humans, too, environmentally induced adaptations could not have positive as well as negative effects on heredity.

After 1910, other eugenicists also accepted the inheritance of acquired characteristics. But because most stressed only the negative effects of environmental influence, the process did not usually entail a rejection of selectionist methods. On the contrary, many assumed that the negative effects could be eliminated only by selection. Peter Bowler presents the example of Kammerer's defender, British zoologist E. W. McBride. McBride was a committed neo-Lamarckian, but he also believed that racial attributes were permanently fixed. In his view, the races had evolved over such a long time that in human terms they were effectively permanent, a view that justified his advocacy of forced sterilization.[14] Kammerer's understanding of the environmental control of heredity differed from other neo-Lamarckians because it incorporated the two other processes that were parts of his larger synthesis. Their addition substantially changed the social implications that he could draw from the inheritance of acquired characteristics.

The first of these was the biology of symbiosis or mutual aid. Kammerer had always been a staunch evolutionist. But throughout his career, he argued against an exclusively competition-based approach to evolution and for symbiosis and cooperation as processes that stand alongside struggle in driving evolution. Kammerer presented his ideas on cooperation and symbiosis in several works in which he cited Darwin's account of human evolution.[15] In *The Descent of Man and Selection in Relation to Sex*, Darwin argued that human morality had evolved from the "social instincts" that animals direct toward kin. But unlike other animals, humans have an automatic altruistic impulse to help even strangers. That impulse had evolved into full-blown morality through the inheritance of acquired characteristics. Kammerer defended Darwin's view, complaining of the "fashionable mania of selectionism" among "neo-Darwinists" who misused Darwin and raised the principle of competition to the "father of all things."[16] They ignored the pervasiveness of cooperation in nature. Noting that he rejected only the exclusivity of competition, not its possibility, Kammerer drew on his vast zoological knowledge to defend the importance of cooperation. He cited numerous examples that illustrated both the pervasiveness and the power of mutual aid and mutual support. His examples included interspecies mutualism in animals—hermit crabs on whom sea anemones rode to obtain leftover food as their toxic barbs stung the crab's potential predators—and those between plants and animals that facilitated nitrogen metabolism—the marine Strudelworm and green algae. Such phenomena were widespread; he viewed even the metabolic interrelationship between oxygen production in green plants and carbon dioxide production in air-breathing animals as symbiotic. Kammerer's insistence on mutual aid and symbiosis as full partners in evolution was motivated partly by his eugenics. It repeatedly appears in his accounts of productive eugenics. "We know from manifold experiences that human nature is not an exception to the character of these large-scale regularities, which are valid for the lower organisms." Beyond the evidence of nature, however, he needed the principle as a foundation for his hopeful view that biology could be applied to society without devaluing the rights of individuals.[17]

The second process that was central to Kammerer's larger synthesis was his reanalysis of the biology of fertility. During the war, Kammerer completed experiments on two species of salamanders that differed in their patterns of reproduction. The fire salamander lays as many as forty eggs at a time; the eggs are deposited in water, they live as larvae that breathe with gills, and their development occurs in water with no attendance or care from the mother. In the black salamander, by contrast, only two young mature at a time, and the two develop as embryos for an extended period inside the mother—her body nourishes them. They are larger than the larvae of the fire salamander, and, when they enter the world, they are born as completely developed little salamanders, far more mature than the larvae of the fire salamander. Two things struck Kammerer when he compared the reproductive patterns of the two species. First, the females of both species started their breeding periods with the same number of fertilized eggs—about forty. But in the black salamander most eggs dissolved and were absorbed as food for the two embryos that developed in the mother. Second, both species usually ended with about the same number of living young: only two. Most of the larger number of eggs laid by the fire salamander eventually failed to develop and died.

To explore the two patterns, Kammerer took female fire salamanders and placed them, not in water, where they normally bred, but on land. The females deposited their eggs as before, but most larvae died in the strange circumstances. But when Kammerer did this repeatedly, he observed that the fire salamanders gradually adapted to the dry terrestrial conditions: they retained their larvae for longer periods in their bodies; the larvae developed to a larger size; and when the young were born, they were more fully developed. Surprisingly, as these changes occurred, the number of larvae developing in female fire salamanders decreased from the original forty to a number more like the two usually produced by black salamanders. Female fire salamanders had accommodated to their changed habitat. Through exposure, they lost the pattern that maximized the quantity of young produced and instead developed one that enhanced the production of fewer, more completely developed young.[18] The pattern was, as Kammerer of course already believed, plastic, not fixed. He then examined how the young of subsequent generations bred. As adults, the young fire salamanders too, became late-bearing and produced fewer more completely developed young. They had inherited the entire pattern of reduced quantity with increased quality that depended upon parental resources being shunted toward the care of fewer young. Kammerer presented the two patterns as a law-like rule describing the inverse relationship between the quantity and quality of offspring in nature. Both were natural patterns; but greater quantity naturally entailed lower quality, understood as less complete development resulting from less care. Kammerer's experiments showed that his law applied within and between species. And most important, a species that began with one pattern could, via the inheritance of acquired characteristics, evolve to exhibit the other.

Many historians of eugenics interpret an emphasis on the quality of attributes to reflect racist or classist views implying that one individual or group is inherently of higher quality than another. Indeed, this was often the case in selectionist eugenics. Kammerer's, however, was a different notion of quality. By defining quality developmentally, he meant the opposite. Quality meant rearing. In adjusting to a changed environment, a species that had produced in quantity could be induced to decrease the quantity and increase the quality of the young produced by investing more resources in caring for fewer individuals. Quality individuals were not selected; they were produced by greater care during development. Importantly, the shift in salamanders occurred with no change in the final number of surviving descendents. This meant that stressing quality did not lead to population decline. As always for Kammerer, this was not a purely scientific exercise. Rather, he believed that the inverse relationship between "offspring quality" and "offspring quantity" applied to human fertility in a way that challenged the prevailing Malthusian views of human population growth.

Mainstream eugenics based the promise of human improvement on the logic of sheer numbers: selection required relatively more births of the "fit" to restore the species. Kammerer argued that this approach ignored the fact that biologically, fertility and the survival of young could be sustained in two ways, both of them natural. Social engineers could improve humanity, not by increasing relative numbers (which Kammerer regarded as a bourgeois view, to which even Darwin had fallen prey) but by increasing the quality of resources and care devoted to raising fewer young. This approach did not assume that some individuals or groups were inherently of superior quality; rather it assumed that by channeling resources toward developmental potential, society could produce greater quality in any group. "Excessive numbers of children are no blessing because they produce more badly developed children whose inborn delicacy is aggravated by the acquired damage of less careful nurturing."[19] Kammerer even implied that by decreasing care, the effort to increase population numbers actually produced the inferiority that it sought to eliminate! His eugenics would instead alter the quality of the environment and of parental care—his examples made it clear that the responsibility should fall to both men and women—and emphasize the value of raising fewer more completely developed individuals in ways that did not sacrifice anyone.

Historical research on eugenics has shown that after World War I, many eugenicists more aggressively stressed the need for negative eugenics. Kammerer too, addressed the war. But his account illustrates how by stressing the quality of environmental resources, productive eugenics would deal differently with the dysgenic effects of war. Selectionists held that "the maintenance of the individual must always be second to the maintenance of the species." But, he wrote: "this renunciation of personal well-being in wartime is not only useless and unproductive, but it produces continual unhappiness well into the nth generation." He was referring to the impact of war on "war-acquired bodily and mental defects in our children and grandchildren" that, via environmental neglect, had reverberated across generations.[20] Improvements in the environment that emphasized the inheritance of acquired characteristics would alter conditions for children that could permanently counteract the negative effects of war. But doing so required an emphasis not on the numbers of "fit" or "unfit," but on resources directed to the personal well-being of all children.

Socialism and the Endocrine Integration

The physiological and developmental combination of Kammerer's central tenets—plasticity, offspring quality via care, and cooperation—formed the basis of productive eugenics. By analogy with the improvements brought to modern life by the application of physics, Kammerer called his approach to human improvement "organic technology." Just as efficient machines harnessed the smallest quantity of energy and converted it to useful work, so too, he argued, organisms must follow the imperative of energy efficiency (*energetisch-ökonomisch Imperativ*). The task of organic technology was to liberate as much as possible the "the slumbering developmental energies (*Entwicklungskräfte*) of the individual and the racial organism and exploit them—in their force and their duration—as completely as possible."[21] Significantly, Kammerer often used the endocrine system to illustrate the developmental power of "organic technology." Modifying the human environment using procedures such as organ transplant, radiation, and hormone replacement would channel developmental plasticity to shape the species in novel ways.

In several works published after 1920, Kammerer stressed the role of the endocrine system in ensuring the plasticity and the cooperation needed to produce these novel adaptations. Here, however, his use of the endocrine system provided more than just a needed physiological mechanism. It was also the ideal way to accommodate his political ends. Kammerer aligned himself with Austromarxism, an Austrian version of socialism that stressed gradual social change over the revolutionary Marxism prominent in Soviet Russia. His socialist vision could not abide the competitiveness and the parasitism that he saw in accounts of evolution based entirely in competition. The union of his politics and his science is clear in his account of how hormones enabled cooperation both within the individual and for the common good. In his dramatic defense of Steinach's rejuvenation procedure, *Rejuvenation and the Prolongation of Human Efficiency*, Kammerer described the endocrine glands of the "polyglandular system," not merely as determinants of bodily proportions and sex drives—or even as the key to somatic induction—but as part of the machinery needed for internal cooperation and symbiosis.[22] *Rejuvenation*'s central subtext is that endocrine physiology is the body's physiological system of teamwork.

In the book's first chapter, Kammerer maintained that aging and even dying are the result of imperfections in metabolism. Normal metabolism can produce "autointoxication," a process of atrophy and deterioration that results from residues left when the "remnants" of metabolism accumulate and eventually clog the tissues. Metabolic purification normally produced, for example, by excretion, respiration, sweating, and alimentation, can naturally eliminate these toxins. But as the organs responsible for these processes—the kidneys, skin, nervous system—fail, toxic impurities build up and take their toll. The eventual results are autointoxication, aging, and even death.[23]

But if the impurities never accumulate, aging need not occur. After listing the endocrine glands, Kammerer argued that the endocrine system was the system that brought about restoration of the integrated functioning whole. The endocrine glands were special organs. They operated, not through ducts but through the entirety of circulation; their "incretions," therefore, influenced any and all areas of the body to which the blood flowed. Because of their connection to the entire body, they could "take care, everywhere," of all distant parts.[24] Unlike the nervous system, which acted in a top-down, hierarchical way, the endocrine system's diffuse blood-borne mode of communication maintained an egalitarian teamwork in which cooperation and balance occurred via decentralized "unification." The separate glands always operated in cooperative balance: "If one [gland] is successfully restored to health, all the others will participate in this benefit."[25] But when the glands deteriorate, the purification process loses its collective coordination. Failure of the naturally purifying effect of metabolism, he said, occurs when "there is a deplorable lack of 'teamwork' between the different organs."[26]

The methods of organic technology could restore "the conditions of those organs and glands which take care of digestion, excretion, respiration, and the general purification of the body."[27] Steinach had shown their developmental potential, and, for Kammerer, endocrine intervention became the outstanding example of how organic technology could harness developmental energies to restore cooperative balance through reinvigorated glandular teamwork. Further, as Steinach's work had shown, the developmental effects of the glands were not adaptive endpoints predetermined in the germ-plasm; instead, they were plastic and sensitive to their environment. Organic technology would draw on that plasticity to rechannel the energies of development by altering the environment in which it occurred. It was this engineered and redirected plasticity that made possible the production of novelty that went beyond the promise of mere repetition by reproductive sorting that was the foundation of selection.

Investing in the plastic energies of development produced novel qualitative changes beyond the preventative measures used in medicine and the selectionist measures of eugenics. Though these might help, selectionist interventions had two major drawbacks. First, they relied on what already existed—attributes that were sometimes the product of damage or chance—and not on the potential to produce novel and more highly evolved (*höherentwickelte*) characteristics. Endocrine manipulation promised true novelty, and for Kammerer that meant progress. It made possible "a eugenic effect in the purest sense, further progress with all avenues open (*Tür und Tor geöffnet*)." Waxing dramatic, Kammerer wrote that the endocrine system determined "the mood and directions of our mental lives; the degree and the special imprint of intelligence, of virtue and vice are to a great extent, if not exclusively, their work. For even likes and dislikes, desire and aversion, that are brought to us by outer experiences first take their path or detour through the endocrine glands, depending on whether they stimulate them to stronger work, more copious secretion, or induce them to decrease their activity."[28] Kammerer's emphasis on offspring quality was an important piece of this synthesis. About children, he wrote, "whoever considers that the differences in the abilities of our schoolchildren can essentially be grounded in variations in the production of their glandular systems . . . whose inner secretions wash the brain and whose amount and composition are sufficient in different degrees and in different ways to produce their cognitive abilities, to that person are opened wide new perspectives on the regular extirpation of mind-body inferiority and the intentional production of gifted capacity using [endocrine] organotherapy that no longer belongs to the most audacious."[29]

Second, mainstream eugenic methods relied on a destructive competitivism that undervalued the cooperative character of nature and thus ignored the compatibility between what was good for the individual and what was good for the group. In 1924, as he challenged the pure selectionism of eugenicists who followed Weismann and saw evolution solely in terms of selection, Kammerer stressed the destructive result: "Class struggle is a veritable struggle for existence: a race with mental weapons, without violence, a bloodless and a positive selection—the survival of the fittest. War is synonymous with negative selection, with the leftover of the weak and halt." From them, no good could come. By contrast, the cooperation and flexibility mediated by endocrine teamwork ensured both collective good and enhanced the quality of all. Drawing on this distributed endocrine teamwork, Kammerer's final integration was socialist to the core. He wrote: "Like socialism, [real Darwinism] is a doctrine of 'upward development' and must concern itself with the masses and not only individuals, or it misses its aim." Supported by natural cooperation, each individual "should fit himself in such a manner as to contribute to the general welfare according to his gifts and receive, in exchange, whatever other members can best afford to give."[30] The excess numbers produced by an emphasis on quantity were usually workers, who became "cannon fodder" for selection. Without an emphasis on the resources needed to produce quality offspring molded by cooperation and without a heredity that could be shaped by improved environments, the working class would never become solid workers whose efforts profited the people. Kammerer's "unadulterated" Darwinism permitted "the metamorphosis and intermingling of races and classes, the power of the struggle for existence, and the even higher power of help to existence. . . . Darwinism is a doctrine of natural, world embracing humaneness."[31]

Nothing about the glands required language like "teamwork," "deplorable," and "duty," though endocrinologists had already described examples of chemical synergy. But for Kammerer, the endocrine glands were the socialist system of the body. With them he could stress cooperation, not competition, as a dimension of physiology as well as of evolution. And he could do so with the very system that, he had argued, insured the plasticity evident in somatic induction. The integrative action of this "polyglandular" system ensured the bottom-up regulation of a distributed cooperating group. In it, "friends and foes, therefore, find themselves within the 'incretory association.' Assistance and antagonism both are contributory . . . in reaching the final goal which is already fixed as the duty of the ductless gland system; i.e. to achieve the harmony of the parts and to build a perfect whole from them."[32]

Kammerer shared this socialist vision with his good friend, geneticist Hugo Iltis. Writing in the Workers' Yearbook of 1926, Iltis clarified the logic of the relationship between socialism and biology, emphasizing the consistency between Darwin and Marx.[33] He defended evolutionary thinking as a foundation of socialism by explaining the two senses of "Darwinism" that were often confused in the public's mind. Darwinism in the narrow sense was Darwin's mechanism of natural selection; Darwinism in the wider sense was the broad evolutionary teaching (*Evolutionslehre*), which, Darwin knew, depended on many processes, and which had given humans their natural place. Reactionary opponents of evolutionary teachings, Iltis argued, had used the debates about natural selection to reject all evolutionary thinking. But these debates had no impact on the wisdom of an evolutionary approach to humanity as a substitute for otherworldly theology and hierarchy. In this sense, there was no "crisis" of Darwinism.

Iltis stressed three new methods that provided "direct" evidence of evolution.[34] The first was the new experimental approach to the inheritance of acquired characteristics. Many experiments—several were Kammerer's—showed that novel attributes first acquired through environmental change could then be inherited. Lamarckism in its new experimental form was, he said, the parallel to Marx's emphasis on the environment as the foundation for history. The second was the rediscovery of Mendel's genetics, a topic on which Iltis was an expert. The third was the mutation theory, which eventually showed the importance of unpredictable genetic change for the processes of species change. All three approaches had their followers, and there was often dissent among them. But Iltis emphasized that evolution could not be explained through any single approach. On the contrary, when the three perspectives were brought together, Darwinism in the broader sense would be among the best established of modern scientific theories.

Iltis also stressed the meaning of evolutionary theory for socialists who aimed to rebuild the order of society. Here, in the year of his death, Kammerer's friend took a page from his rhetoric. It was the page of hope; and for Iltis too, that hope was based in the validity of the inheritance of acquired characteristics. Special creation, or any theory of unchangeability in nature, would always serve as the basis for a conservative worldview—one that maintained

(Iltis quoted the hypothetical conservative): "So was it always, and so will it remain." To this, the hypothetical socialist must respond: "What once was can and must become different." The socialists' reforms would be produced by a Darwinism that accepted the inheritance of acquired characteristics, and it was understandable, Iltis wrote, that "reactionary elements" would be found among the ranks of the anti-Lamarckians. "With experience we could establish a developmental direction in organic nature that coincidentally parallels that which from our [socialist] standpoint we call 'more elevated (*höherhinauf*).'" This multicausal evolutionary synthesis, Iltis said, had special meaning for the struggling socialist who sought to build "the ideological superstructure of a socialist communal order. . . . It gives us the optimism needed by each person who struggles for something new."[35]

Race and the Critique of Selectionist Eugenics

In the late 1910s, as Kammerer was developing his ideas on the endocrine basis of somatic induction, the intensity of both his hopeful idealism and, eventually, his deep despair increased. There were three major reasons for this bipolar increase. The first came with his realization of the value of endocrinology, especially the interstitial cells, as the mechanism of somatic induction and the key to harnessing the energies of development. The second is related to the ways in which World War I changed the flavor of eugenics. The horrible death and destruction produced by the war increased the perceived threat to humanity. And this produced greater emphasis on negative eugenics, which, by 1920, was even more strongly linked to the idea of race and to race science. Many of the most "fit" had been killed, the logic went, leaving the "weak," the "inferior," and the "cowardly" as the basis on which to rebuild humanity. In response, Central European eugenicists increased their emphasis on negative selective breeding to assure the ascendance of the "fit." The third lay with changes in the international scientific community's understanding of heredity. The American school led by T. H. Morgan increasingly challenged the inheritance of acquired characteristics.[36] Morgan's Mendelian-chromosome theory of heredity was Weismannist to its core, and by 1920, more and more research suggested that Weismann could be right. Early in the century, the changeability of race had been an issue in debates about the validity of the inheritance of acquired characteristics.[37] That too, was changing, and the combination of events terrified Kammerer. The war provided what he saw as a faulty social justification for an alliance between negative eugenics and the Weismannist base of genetics. That alliance was fed by the sense that an important new science—genetics—had emerged that rejected the inheritance of acquired characteristics. Kammerer saw their union as scientifically flawed, unethical, and, if unchecked, disastrous for humanity.

Even some fellow socialists who had rejected forced measures based on negative selection, began to advocate them in the early 1920s.[38] At war's end, Kammerer became convinced that the basis for this convergence was the racism growing within science itself.

As a result, Kammerer became more outspoken in his criticism of the racist base of selectionist eugenics. His alternative, as I have shown, was based in cooperation and the productive power of development to alter heredity and generate true organic progress. Many of his works laid out the argument. But a few also began to condemn mainstream eugenics as much on racist as on scientific grounds. In them, Kammerer promoted his alternative as an ethical challenge to the racism of selectionist eugenics. In a small book entitled *Lebensbeherrschung: Grundsteinlegung zur organischen Technik* (Control of life: foundation of an organic technology), Kammerer supported the value of eugenics. His initial summary seems to place him squarely within the classical camp. He included, for example, a discussion of Mendel's rules of selective breeding and their value to humanity; and he reviewed—and seemed to endorse—the selective measures that had recently been made law in the United States. He even noted that since the war, Europe had become a "breeding station for inferior human materials."[39] But Kammerer continued by distinguishing two very different kinds of eugenics.

The eugenics movement was, he said, already divided into two camps. One consisted of the followers of Weismann's theory of a sequestered germ-plasm; the second accepted the inheritance of acquired characteristics. The former, he said, saw gene selection (both positive and negative) as the only mechanism of evolution and organic change; the latter sought other modes of change. In contrasting them, Kammerer's tone changed. He not only rejected the first view, he criticized the motives of those who promoted it. They were racists. They had no regard for the processes of development (he labeled them *entwicklungsfeindlich*—enemies of development). Because they ignored the organic origins vested in environmental influence, he thought they should be called "anti-geneticists." This camp believed that racial characteristics were due to irreconcilable hereditary differences rigidly inherent to a group, a view that Kammerer saw as a silly form of preformationism. More strongly, he accused this camp of being the home of "exaggerated, dark nationalism, of short-sighted chauvinism." It was the domain of "racial fanatics, who in the dominance of only one race (the blue-eyed blond) expected the healing of the world and desired the destruction of all other races."[40]

The second camp recognized the role of the environment in heredity; it acknowledged, he said, the potential of all individuals to be transformed into the improved *Anlage* of the species. For this reason, unlike the other, this camp valued the worth of all individuals. Further, Kammerer noted, for this camp the struggle for existence would be made milder by the countervailing influence of mutual aid, by "help in existence (*Hilfe im Dasein*)."[41] The "fountains" that would create effective eugenics would include "theories of variation, heredity, and descent with their observations of struggle *and* aid in existence (*Kampf und Hilfe im Dasein*), of selection, breeding, *and* adaptation."[42] Cooperation could even dominate competition in its application to humanity. So, though selection operated in nature, more humane and egalitarian alternatives were available; and on scientific, ethical, and progressive grounds, Kammerer preferred them. But this reduction in competitive struggle could occur only once the equal worth of all human beings and a deep sense of responsibility for the nurturing institutions of society were instilled in everyone. With selectionist thinking replaced by environmental reform and education, the inheritance of acquired characteristics would do this without recourse to either "positive" or "negative" restrictions on procreation, which Kammerer saw as inhumane and racist.

Kammerer had already expressed his views on race in 1918, when he summarized the thinking of his contemporary Friedrich Hertz. In the recent edition of a work originally entitled *Moderne Rassentheorien* (Modern race theory), Hertz challenged the existence of pure races, arguing that both race mixing and outer living conditions completely blurred any racial differences. Neither the skull measurements nor, especially, the psychological attributes that race theorists invented to support moral and mental inequality existed. For Hertz, the race concept had only social grounding. Kammerer was flattered when a Catholic educator, critical of Hertz's views, attributed Hertz's ideas to him and, in the process, called Kammerer a pacifist, a monist, and a socialist. Endorsing Hertz's ideas and approving of each of these "honorable" labels, Kammerer noted that only minor differences distinguished his views from Hertz's. While Hertz treated race as a social ideology, Kammerer considered it something real, but "relative," "fluid," and, therefore, "transient." It is not clear whether Kammerer was citing the critic, Benno Immendörfer, or expressing his own thoughts when he defined a socialist as "one who promotes the same rights and obligations to all." Whichever, Kammerer clearly approved of those, including himself, who he satirically labeled "the scum of humanity" because they fought for such egalitarian goals.[43]

Kammerer's criticisms of European selectionist eugenics and his socialist logic did not prevent him from trying to persuade eugenicists in capitalist America of the value of the inheritance of acquired characteristics. In 1923, Davenport, the guiding force in American eugenics, wrote to Kammerer inviting him to lecture at the Cold Spring Harbor Laboratories on Long Island, headquarters of the American eugenics movement. Davenport had heard that Kammerer was in America, and in the invitation he noted that Kammerer "would be interested to see the work that is going on here." Kammerer accepted and inquired as to why Davenport had not pursued a plan, which Davenport had described to Przibram during an earlier visit to the Vivarium in Vienna, to replicate Kammerer's salamander work. When Davenport replied that the effort had stalled because the right species were not available, Kammerer offered to provide the needed animals. Noting that others had already replicated the studies, Kammerer took the opportunity to stress how important it was that scientists at Cold Spring Harbor do "experiments as to the inheritance of acquired characteristics . . . be they my own or the experiments of others."[44] Kammerer's American tour was contentious; Morgan had refused to be on the scientific organizing committee for the visit, and we can only imagine Raymond Pearl's reaction. Nonetheless, even after he had decried the changes in European eugenics (which drew on the American model), Kammerer held out hope that, despite their rapid acceptance of Weismann and Mendel, American scientists might be open to another approach to human improvement.

Kammerer wrongly perceived America as less inclined to racism than was Central Europe. In 1926, he cited Franz Boas when he wrote that the United States was a showplace for how foreign "races" could be amalgamated into a blended people and absorbed into a uniquely American nation. Following Boas, he stressed that American soil had changed its immigrants. The environment had produced a convergence of physical and mental differences, and the longer immigrants remained on that soil, the stronger the transformations were. The result was not a newly selected "Yankee-Type," but a "bidirectional assimilation of individual features," a conclusion borne out by Boas's measurements on the descendents of immigrants. Kammerer stressed that though the environment would be unlikely to erase all racial differences—immigrants had brought conservative aspects of their home cultures with them—it would be powerful enough to produce adaptations that furthered "an assimilation in the general molding (*Gesamtgepräge*)."[45] The result, he said, would be a triumph of the feeling that all humanity was composed of brothers.

Kammerer knew of the opposition; he acknowledged the racist views of Madison Grant, the Ku Klux Klan, and Zionists such as Charles Elliot of Harvard. And in advocating the value of Steinach's procedure in the United States, Kammerer risked compromising his own argument against selection. Steinach's rejuvenation procedure was, in effect, sterilization. Kammerer was advocating its therapeutic effects just as sterilization began to dominate the logic of American eugenics. In 1925, as he reiterated that heredity must be designated "an inner-secretory process," he described the selectionist measures used in the United States. Reversing himself, he said that they could, via the glandular system, reduce the heritability of the criminal instinct and even restore health to the affected person.[46] In his 1924 book, he had pointedly held the opposite view, arguing against the poverty of (sterilization-based) selection as against the novelty and progress possible when environmental change alters heredity. The contradiction reveals that Kammerer, too, would consider inhumane measures. His cultural naiveté about race in the United States (Boas's antiracial arguments were censured by the American Anthropological Association before 1920), his passion for the endocrine alternative, and his desperation to find a culture less based in racial hierarchy than his own made the United States seem more benign than it was. By 1923, Kammerer was eager to leave Austria, and he needed the United States to be a promised land of racial tolerance. That in the 1920s—a period in which the Supreme Count sanctioned forced sterilization and Congress passed the most restrictive immigration law in its history—the United States could appear to be an experiment in racial blending and balance speaks to the idealism guiding Kammerer's science. Yet even as he acknowledged the benefits of sterilization, he stressed selection's conservative effect. It could remove burdens, but it could not create good. Kammerer sought more. Members of all races were part of the general "racial worth" of humanity; each person had the potential to improve the species. By environmentally harnessing the energies of development, that potential could be realized. In the end, it should not be necessary to discard anyone.

His idealism reached still deeper. Because it could instill in the species greater dominance of the impulse for mutual aid, organic improvement did more than just mold quality individuals. It also promoted the evolution of a biologically based sense of ethical responsibility, a hereditary *Gesamtwille*, an impulsive drive for mutual aid that was just as much a result of education as was any individual ethic.[47] As we enact ethical measures and become more and more conscious of our obligation to the quality of future citizens, that ethic would become hereditary. Organic technology could expand the welfare of the person "into a social technology for the well-being of a generation. . . . Living conditions should . . . be raised in all segments of the population, because only in this way do those racial qualities manifest themselves—first prepared as personality attributes—that in an appropriately milder struggle for existence become the possession of a generation."[48] Increasingly, Kammerer could interpret the union of nature and culture as a product of two principles—symbiosis and the inheritance of acquired characteristics—that in their mutual support would even produce a biologically more ethical humanity. Kammerer's writings had often referred to "race apostles" such as Grant and Chamberlain. In the postwar years, Kammerer positioned himself as an anti-race apostle.

Kammerer's Ethical Circle: Austrian Monism and Developmental Value

Productive eugenics was eugenics that used the openness of evolution and development to direct environmental input and mold novel phenotypes. It was a developmental bioengineering of mutual aid that relied on a pluralistic view of evolution to achieve an immanent collective ethic. Kammerer had devoted his career to the evolutionary principles on which the integration depended. That integration could, with Steinach's endocrinology, instill a deeper morality if humanity modified the environment to "touch the sensitive mechanism of the germinal systems" and make "our feeling of responsibility . . . different, incomparably greater . . . not through outer tradition handed down in oral and written form from generation to generation; no, in addition in a much more deep-seated, because organic, way, through the sensitive nervous and chemical (innersecretory) mechanism."[49] Kammerer's move to endocrinology and his social commitments were of a piece; and there is, I think, no better way to see how completely his values infused his science than to place him in the context of the Central European monist movement. His involvement with Austrian monism illustrates that his egalitarian, cooperative view of humanity and his objections to race science were not produced by an isolated voice. Rather, they were embedded in a larger circle of scientists, socialists, reformers, and ethicists whose views pointedly challenged the growing Central European reliance on racial hierarchy. And for them Kammerer's science added the critical scientific piece.

Founded in 1906 by the German biologist Ernst Haeckel, the monist movement defined a secular worldview anchored in evolutionary biology and dedicated to developing a naturalistic approach to humanity.[50] It rejected supernatural and spiritualist accounts of human existence based in traditional religion and aimed to replace them with a unified view of society and nature, mind and body—a oneness of humanity and the natural world that depicted nature and society as governed by similar laws. Especially as represented by Haeckel, historians have often focused on the negative aspects of monism. Indeed, monism shared eugenic ends; it was not immune to the biases common to much early twentieth-century eugenics. The standards of accomplishment that monists sought were often dictated by paternalistic white, middle-class values: workers and members of other races had to be helped by Western science to achieve inner improvement. Further, monists courted paradox when they privileged biology as the source of these laws: in a truly unified conception of life, why would one domain dictate the rest? But most founders agreed that, anchored in the progress made possible through a modern understanding of ecology and evolution, monism offered a vision of improvement that placed humanity squarely within nature.[51]

The monism movement grew quickly in Germany, Austria, and Switzerland. By 1911, the year of the first international monist congress in Hamburg, there were five thousand members and chapters in thirty-six German cities as well as in Vienna, Zurich, and Geneva. Kammerer's connections with Haeckel's student Richard Semon placed him within the inner circle. And in 1911, with Rudolf Goldscheid and Wilhelm Börner, Kammerer co-founded the Austrian Monist League

(*Monistenbund in Österreich*). Austrian monism was closely linked to socialism and to the reform traditions growing around the Vienna Circle, the philosophers and physicists who sought to reform science on inspiration from Ernst Mach. Friedrich Stadler writes: "Monism was praised by its Austrian representatives as an instrument for the emancipation of the masses out of their dogmatic slumbers. With the help of monism they attempted to fight against and bury each form of dualistic thinking (anthroposophical, intuitive, spiritualistic, idealistic)."[52]

Austrian monism's quest to improve human life through an all-encompassing naturalism required the validity of two things that were the foundation of Kammerer's science. The first was demonstration of the role of outer influences in producing novel adaptations in the species—somatic induction in inheritance of acquired characteristics. Monism had been founded on the concept, and after Semon died, Kammerer's rejection of what he considered metaphysical dualism in heredity became a key dimension of Austrian monism.[53] The second was the recognition of the evolutionary power of mutual aid. The war had shown that unbridled competition produced unprecedented destruction. Kammerer believed that it was no wonder that the public rejected ideas of evolution based solely on competition; they had seen the result. "Translated to humanity, this biological struggle means primitive competitive discontent about daily bread, enraged racial hate, and, when it is most terrible, political war!" But if evolutionary theory were also linked to a positive impulse for mutual aid, monism could counter the biology of perpetual struggle and show people a world of humane values that was just as consistent with science and nature. Kammerer wrote: "When it is seen in terms of general symbiosis (*gegenseitige Daseinshilfe*), great marketing power inheres in evolutionary theory to help turn the masses to a healthier world view that is free of mental contradictions."[54]

Though the tenets of Austrian monism were shaped by Kammerer's science, the dynamic, progressive, and ever-developing processes of life did not automatically dictate positive and productive change. On the contrary, values inhered not in biological processes but in a society inclined to shape them humanely. Nature showed a spark, but environmental change could also perpetuate evil through "neglect, hateful quarrels, the repression of knowledge, and job slavery (*Berufssklaverei*)."[55] It would be social engineering and monist reform that gave the scientific principles their ethical center. And the implementation of Kammerer's progressive values was almost certainly shaped by his Brno-Prague-Vienna circle of anti-racial activists, who eventually challenged Nazi racial ideology.[56] The group was a diverse assemblage of biologists, sociologists, philosophers, and social critics, many of whom were fellow monists, and Kammerer looked to them to hone the values that framed his biology. In Vienna, the philosopher Friedrich Jodl was their guide.[57]

It was Jodl who first attempted to establish the movement on high ethical and moral ground. Because it challenged the hegemony of religion, monism was open to the criticism that it undermined traditional human morality. For this reason, a social and political ethic had to become part of its secular view. Jodl framed that ethic in an address given at the first international monist congress. In it he summarized a one-word answer to how the monist movement should perceive its relation to contemporary cultural problems: ethics. Modern monism had been a product of natural science, "a child of . . . evolutionary teaching." Because it embraced evolution and challenged clericalism, he said, monism was often seen as a form of atheism: a worldview that offered no vision of moral value. Jodl warned that unless it incorporated an ethical perspective, monism would never have a lasting cultural influence and might even become a danger. Monism's task was, therefore, to develop an ethics that was consistent with its scientific foundation and with the oneness of nature/culture, mind/body—a oneness "whose full meaning had perhaps not yet been appreciated."[58]

Accomplishing this task required two things. First, morality without the imposition of religious or transcendental concepts must be more than just a theoretical possibility; instead, it must grow strong roots in society, roots that could sustain monism not just among intellectuals but also among citizens in the broader culture. Second, any set of moral values must come, not from laws brought down to humanity from outside but "as an evolutionary product (*Entwicklungsprodukt*) of the living exchange between the individual and society, as the fruit of a balance between the individual will and the common will." When morals and a scientific ethic are understood as a union of the natural, the moral, and the cultural, then, Jodl maintained, ethics would be more than an intrusion (*Einbruch*) or add-on to natural science. It would instead become the correct way to go forward from animal to man, from "the struggle for existence to the struggle for rights; from bestiality and brutality to humanity."[59] Such an ethic would not be monism's enemy but its crowning glory. In the practical advocacy pursued by monists, these values were seconded by neurologist August Forel and expressed by him as a pervasive pacifism, as the quest for equal rights for men and women (including suffrage), as a profound concern with the well-being of youth, including public education and sexual reform, and, paradoxically in hindsight, as eugenics and race hygiene.[60]

We do not know whether Kammerer heard Jodl's lecture. He may have. But he did not need to. Kammerer already looked to Jodl for moral guidance. Jodl had come from Prague to the University of Vienna in 1896 as professor of philosophy. He was Kammerer's philosophy professor, and the two men continued to interact well after Kammerer's university years. On Jodl's death in 1914, Kammerer wrote Jodl's widow describing his admiration for the older man. Jodl's influence was, he said, so profound that it was still with him, giving Kammerer his broader philosophical view.[61] Jodl's argument for ethics as an immanent "evolutionary product of the living exchange" between the individual and society almost certainly fed Kammerer's symbiotic solution to the integration of biology and society. Symbiosis as a full partner in evolution was a response to Jodl's call for an ethic that began as a cultural value but became a dimension of biology, one imbedded in individual heredity and productive of both collective values and the respect for individual potential. Kammerer's understanding of upward development (*Höherentwicklung*) referred, not to an inevitable teleological process inherent within evolution but to the socially integrated and natural development of a collective ethic that was stressed by Jodl and would be enhanced when the seed of symbiosis, present in nature, was expanded by social engineering to guide the induction of an automatic ethical impulse.

That Kammerer's synthesis and his advocacy of the inheritance of acquired characteristics continued to shape the Viennese ethical community after Jodl's death—indeed after Kammerer's death—is indicated by Wilhelm Börner's copy of Kammerer's *Neuvererbung* (see note 8 above). Börner studied philosophy with Jodl; he helped found the Austrian Monist League, and after Jodl's death he led the Viennese Ethical Union (*Wiener ethische Gemeinde*). Though not a scientist, Börner would have known both of the advances in Weismannist interpretations of heredity and of the need for a physiological account of use inheritance. And he approved of Kammerer's physiological solution. In his copy of Kammerer's 1925 book, Börner made special note of Kammerer's use of endocrinology to explain how the environment might alter heredity. He placed distinct marks in the margins next to those sections in which Kammerer explained how the endocrine glands could regulate heredity. The marks were especially dense and of unique form where Kammerer wrote "Without the mechanism of the endocrine glands, newly acquired characteristics could not become incorporated either into the body of the directly affected individual or into his germ-plasm."[62] Börner used the same marks where Kammerer described the ethical implications of productive eugenics for a more humane approach to racial difference. Apparently the two men knew one another. On the tenth anniversary of Kammerer's death, Börner is quoted in the Viennese newspaper, *Die Neue Freie Presse,* describing Kammerer as an excellent human being for whom the profound utility of natural science to improve humanity was more important than his own research.[63] Börner shows how important Kammerer's science was for monism's ethical foundation. If sustained cooperation and flexibility were biological processes maintained by an increasingly well-understood endocrine mechanism, urban reform and cooperation could, after all, unite individuals of all races and classes. They could instill the ethical impulse that Jodl had stressed in 1911, and that, by 1925, could be understood scientifically in terms of Kammerer's endocrine hypothesis of somatic induction.

Economies of Ethical Value

Jodl provided the ethical theory; but Kammerer's view of the practical implications of the monistic worldview for social order, economic progress, and political change were also those of the founder of scientific sociology in Vienna, Rudolf Goldscheid. Even before the war, Goldscheid supported the importance of science in an aggressive activism that stressed the obligation of science to serve. After World War I, as the head of the Austrian Monist League, Goldscheid pursued the goals of Austrian monism on strongly evolutionary, socialist, and economic grounds.

Goldscheid's concepts of human economy (*Menschenökonomie*) and organic capital linked progress to work, welfare, and socialism. They were built on an interface that unified biology, politics, economics, and social reform along monist lines. In both concepts, the economic value of compassionate care was a matter of ethics as well as of monetary value.[64] As such, they fused the ambiguity of the two meanings of "value": the moral and the economic. Some thinkers used the concept of human economy to support traditional eugenics by underscoring the economic disadvantages of caring for "inferior types." In his analysis of eugenics and socialism in Sweden, for example, Alberto Spektorowski argues that Swedish socialists sought a "higher quality national stock that contributed to productivity," creating a "welfare state of the fittest" that pitted national productivity against the traditional moral values of socialists. As a result, Swedish socialists condoned sterilization.[65] But Goldscheid's notion of value intended the opposite. Goldscheid stressed the need for compassion and the protection of the work force from exploitation as joint pillars of a sound economy. His concept of developmental quality (*Entwicklungswerttheorie*) taught that extending humanitarian and ethical values to human development avoided the waste of neglect, disease, and debility. Instead, he sought to instill an ethic that stressed the value of human potential as "organic capital" in a "developmental economy (*Entwicklungsökonomie*)" that saw mental competence as its greatest economic asset. He wrote: "If we put together the direct and in particular the indirect costs of innumerable occupational diseases, premature exhaustion of the labor force, unsanitary women's work, cheap child labor, insufficient school education, high infant mortality, miscarriages and stillbirths, widespread diseases like tuberculosis and syphilis, alcoholism and prostitution, and if we take into account what the wasteful exploitation, which violently tends to push the envelope of productivity, results in for society, we will realize at once: we are running our household most uneconomically, indulging in a fiction of productivity, and we are paying dearly for this mere illusion."[66] Like Kammerer's, his too was a developmental approach in which productivity and morality were not incompatible. Ethical value was a part of economic value. Mobilizing resources early enough in an individual's life made the scalpel superfluous. In a fashion characteristic of Austrian monism, the individual and the collective were fused—their union required what Goldscheid described as the "collective understanding of an individual (*individuellen Sozialvernunft*)." The economic goal should always be "the development of individual potential with the simultaneous enhancement of collective understanding."[67]

Goldscheid was a sociologist; but his monistic foundation shaped his view of society in ways that required Kammerer's biology. He agreed that Weismann's absolute separation of soma and germ-plasm was wrong. Instead, he presumed a union of the inner and the outer, which viewed the soma as the ever-present mediator between the outer world and the germ cells: "There is a constant and ceaseless alternating current (*Wechselstrom*) that flows from outside to inside, from inside again to outer milieu. What yesterday was still milieu, is today inner disposition (*Anlage*), what today is still inner disposition, is tomorrow milieu." Goldscheid linked this dynamic current to Claude Bernard's physiology of constancy and homeostasis. And with it, he said, we can understand "the economy of relief (*Entlastung*) and burden (*Belastung*) . . . in the service of advanced evolution (*Höherentwicklung*)."[68]

Like Kammerer, Goldscheid rejected the short-sighted and "bloodthirsty" selectionism of Social Darwinism in favor of a view that privileged the environment.[69] And he preceded Kammerer in challenging the Malthusian assumptions underlying the stress on offspring quantity. Society should decrease, not increase, birth rates, investing resources in ways that recognized that human beings, not objects, embodied the greatest value. Goldscheid's neo-Lamarckism, his rejection of Malthusianism, and his monistic assessment of the fluidity of inner and outer influences all emphasized the power of development guided by environmental change. Organic flexibility was necessary to the reform promise that linked Austrian socialism and Austrian monism. Following Jodl, Goldscheid extended the economic implications of that promise to pacifism, to workers' rights, and to the equal rights of women. He even argued that the basis of the organic damage often associated with inferiority could be found, not in an isolated germ-plasm but in negative environments that had taken their toll on human ability and on the human disposition that was needed for the foundation of a productive society. Organic capital did not undermine the quality of "inferiors"; on the contrary, it championed the full developmental potential of those who had been downtrodden.

Goldscheid and Kammerer were friends; Kammerer was thirty-one years old when the two men co-chaired the Austrian Monist League. Kammerer cited Goldscheid often in his accounts of the ethical implications of productive eugenics, and he regularly linked his biological views to Goldscheid's analyses of social ills.[70] Goldscheid's terms *Entlastung, Belastung, Gesamtwille*, and *Höherentwicklung* were among those on which Kammerer built his vision of organic technology. Even in his description of the importance of emphasizing offspring quality, Kammerer cited Goldscheid. As a monist would hope, Kammerer simply strove to show that the parallel principle worked in nature, too. Fewer, qualitatively healthier, better-nurtured individuals rather than greater numbers were features of both a humane society and salamander biology.

The two men dealt with the application of similar ideas to the distinct domains that linked their views. This may reflect their ongoing exchange of ideas. For this reason, it is often difficult to draw clear lines between their ideas. But on one dimension Kammerer clearly led: Goldscheid's views accepted—even required—Kammerer's neo-Lamarckism. The entire social structure stood or fell on the validity of the inheritance of acquired characteristics understood in physiological terms. It was the means to achieve organic change that could be actively shaped by environmental circumstance and by social values. Goldscheid's term *Höherentwicklung* is sometimes translated as "evolution" or "upward development." But these usages can imply a progressive idealism, suggesting inevitable utopian progress. For Goldscheid and Kammerer, it was none of these. Left to their own devices, things could go either way. But, when guided by *Entwicklungswerttheorie* and Jodl's stress on ethics, humans were in control of their own biological fates.[71] To promote change in ways that valued all people, the concept had to be anchored in a modern treatment of somatic induction and in a fusion of the sciences (natural, mental, and social) that rejected the biological separation of nature and nurture, germ and soma.

In 1926, the year he died, Kammerer applied the logic of monism to the science of heredity. He published a monist critique of the new genetics in which he described the dualisms embedded in Weismann's concept of separation and in Mendel's laws of heredity.[72] Adherents of these theories accepted four kinds of dualism. The first was the separation of germ and soma. This idea, he believed, amounted to a scientific version of the religious dualism inherent in the theological separation of mind and body. The second was a separation between organism and environment. This, he said, flew in the face of considerable scientific evidence, especially in vertebrates and plants, showing that under some circumstances, the soma "had the ability to stand in for the germ cells." The third was a dualism of person and generation. The inheritance of acquired characteristics showed that individually acquired adaptive changes in parents could, under certain conditions, be sustained over several generations. Their persistence in later generations established continuity of person and generation and of the ethics that linked them. The fourth was a dualism between human and animal. Several authors had complained that although there might be evidence for the inheritance of acquired characteristics in animals, there was no good demonstration of it in human beings. Monism, of course, assumed no fundamental difference between humans and animals. Based in the evolutionary continuity of human phylogeny, Kammerer assumed, "The Human is an animal, a mammal. What is accurate for animals must also be granted to humans."[73]

With Jodl's ethics, the emphasis on symbiosis and on developmental quality, and an acceptance of Goldscheid's social vision, all linked by Kammerer's endocrine solution to the problem of somatic induction, the concept that dominated Kammerer's life—the inheritance of acquired characteristics—provided a creative reshaping of humanity that controlled selection could not accomplish. For Kammerer, their fusion instilled an ethic without which any attempt at eugenics would become a social disaster. He had cautioned that selectionist eugenics had become the home of "racial fanatics who in the dominance of only one race (the blue-eyed blond) expected the healing of the world and desired the destruction of all other races."[74] Especially in Central Europe, he warned, acceptance of the inheritance of acquired characteristics was challenged because it had become "detrimental to the ideology of nationalist self-consciousness (*dem völkischen Selbstbewußtsein*)," because it promised to dissolve racial boundaries, to blur national contrasts, and to discredit the notion that one race would advance only at the expense of others.[75]

This notion was just what Kammerer's productive eugenics and the ethical circle that supported it aimed to avoid. In 1925, the year before he died, Kammerer again described what his alternative eugenics sought. We need, he wrote, "a more beautiful, more humane way than that which the fanatic race apostle . . . always proposes. . . . Through racial hate and [the] extermination of races . . . we will not save human society from degeneration, never reach ever higher accomplishments, ever higher objectives." Instead, "a new ethic, a new form of knowledge has become necessary: a generational or racial conscience—but not that in the outdated barbaric form in which a single preexistent and prepotent race wants to ensure dominance."[76] In Kammerer's mind and those of his monist colleagues, this new, more humane ethic was possible only if the inheritance of acquired characteristics were a valid process of heredity.

PART II

Reform Eugenics

6

Heredity, Glands, and Human Constitutions

Paul Kammerer was convinced that his three converging principles, the inheritance of acquired characteristics, mutual aid as a principle of evolution, and offspring quality as a principle of fertility could transform humanity. Modifying the environment in humane ways that focused on the young would improve the species, while respecting human dignity and human rights. Quality nurturance could even instill a new ethic—a species-wide drive to help that would elevate humanity to a deeper morality. He had what he believed to be a solution to the troubled state of Central Europe at the end of the Great War. But Kammerer was a relative nobody; though an accomplished scientist and technician, he worked as an assistant at a laboratory on the outside of Vienna's elite academic structure. And he had powerful detractors. What could he do to implement his vision? Partly through his connections with Steinach and with Goldscheid, Kammerer acted to influence the one man he knew who was in a position of power and who would be sympathetic to his concepts, but with whom he had had little prior contact: the Viennese anatomist and politician, Professor Julius Tandler.

In 1917, Kammerer wrote to Tandler, offering to serve in the newly founded imperial ministry of public welfare. Kammerer rightly assumed that Tandler would play a role in organizing the ministry, possibly even direct it, and he hoped to get involved so as to apply his biological vision to the public good.[1] While the monarchy held, Tandler was not appointed to head the ministry. But only a year later, in November 1918, the Emperor Karl, who had just assumed the throne after the death of the long-ruling Emperor Franz-Joseph, abdicated, and the Habsburg Empire, which in one form or another had ruled for over seven hundred years, fell. The socialists took power in 1918 in the coalition government that led the new Austrian First Republic.

Tandler briefly served that government as the national undersecretary of public health. But the coalition fell in only two years, leaving the socialists without much influence in the national government. After the change in government, Tandler turned his efforts to the city of Vienna, where the socialists maintained strong control. In Vienna, Tandler became municipal councilor of health and social welfare. In this position, he wielded great power to shape the reforms that affected the health and well-being of families and youth in a capital city that had been devastated by war. He brought to the task a scientific framework that was strikingly similar to the one Kammerer was developing, but which, in Tandler, was accompanied by an elite scientific reputation, prestige, and position.[2] He had been trained by the premier anatomists of Central Europe, and following the early death of his mentor, Emil Zuckerkandl, he was called to Zuckerkandl's chair of anatomy at the University of Vienna. By 1914, he was dean of the medical faculty in one of the premier medical schools of Europe. As an academic physician and a scientist who was committed to health and social welfare, Tandler was in an excellent position to influence municipal social policy, and he did. He eventually spearheaded the fight against tuberculosis, rampant after the war; he reorganized, reformed, and founded hospitals; and he worked tirelessly for youth reform. By the mid-1920s, he was one of the most respected reform experts in a city that was filled with medical expertise.

Though Tandler has received much historical attention—some of it quite critical—for the health reforms he implemented in Vienna after the war, his impact as a reformer and policy maker has not been effectively integrated with his science. As a medical scientist first and a politician engaged with public health second, his attitudes toward reform were deeply influenced by his scientific work and the assumptions underlying it. In this and the following chapters, I situate the reforms he implemented in Vienna in the context of his scientific views, their overlap with Kammerer's turn to endocrinology, and Tandler's political position during the postwar period, a period in which selectionist eugenics emerged in the new Austria. The socialist's reforms were steeped in political, scientific, and social tensions. Sometimes, Tandler felt those tensions personally through, for example, antisemitic attacks that followed his rise to prominence. They altered the way in which he presented his science, and they affected his advocacy for the reforms that were shaped by it.

Beyond self-interest, which probably played a role, Kammerer's offer of help clearly sought to leverage Tandler's power to effect change. With the socialists in power, the potential to influence reform must have seemed greater than it had ever been. But why did he believe that Tandler would be sympathetic to his science and the reform vision Kammerer derived from it? To what extent were Tandler's social policies—particularly his vision of eugenics—informed by Kammerer's productive eugenics? I argue that Kammerer's presumption rested on solid ground. The two men's scientific views overlapped in important ways; and several aspects of Tandler's science gave Kammerer good reason to believe that he might be positively disposed to productive eugenics. They include Tandler's dynamic, physiological, and developmental view of anatomy in relation to the individual; his early pioneering work in reproductive endocrinology, most especially his advocacy of the secretory function of the interstitial cells; and his career-long acceptance of the inheritance of acquired characteristics, which played a key role in Tandler's framing of constitutional medicine. Though the two men's views also diverged at times, on these points there was nearly complete overlap, an overlap that Kammerer hoped would lead Tandler to accept his help in enacting a version of productive eugenics in Vienna.

Tandler's Medical Career: The Viennese Context

"What is culture, if not to wheedle from the coarse material of life, by art and love, its finest, its most delicate, its most subtle qualities?" Writer Stefan Zweig called his native city of Vienna an epicurean city, a city immersed in culture and consciously dedicated to the enjoyment of the arts even more than of food. By 1900, the capital of the Austro-Hungarian Empire was rivaled only by Paris in relishing in pleasure, especially pleasure through art and culture.[3] Ironically, many Jews, like Zweig, helped shape the quality and impact of that culture, and they identified strongly with it. But Zweig's picture reflects the illusion of a Vienna built by an earlier promise of liberal reform. That promise was strong after the formation of the Austro-Hungarian Empire, which, following a break with Prussia, was established as a constitutional dual monarchy in 1867. Many liberal reforms were implemented in the process, but the situation changed dramatically after 1879. In that year, the reforms began to be replaced by rising nationalist conflicts in what had long been a multinational empire. Growing tensions were complicated by the rise in Austria of a pan-Germanist nationalism whose supporters looked to Germany to negotiate the diversity of the Habsburg Empire. Historian Steven Beller notes that Zweig's nostalgic view of "Viennese Jewry as fitting right in to this world" was a bit of retrospective wish fulfillment.[4]

After 1910, what Zweig calls the delusion of "touching confidence" that had been wrought by years of peace, professionalization, and increasing middle-class Jewish wealth was precariously positioned on the verge of collapse. That collapse came with a war of unimaginable proportions, ending just as Tandler began his political career. World War I and the political and economic turmoil it brought destroyed the empire and the city. At war's end, Zweig's swelling pride was overshadowed by poverty, hunger, disease—tuberculosis was called the Viennese disease—death, wide-ranging homelessness, and virulent anti-semitism. The once sparkling culture of Vienna, still the capital, was now psychologically and politically separated from the small agrarian state that it governed. The war and the empire's dissolution at the hands of the victors produced a national crisis of identity compounded in Vienna by insecurity, hunger, disease, economic unrest, and chaos. Yet it was in this period, between 1880 and 1920, that Tandler rose from modest Jewish origins, sometimes in extreme poverty, to the leadership of medical Vienna. He grew up, matured, and eventually took power amid the transition from hope and prosperity to disease and desperation.

Like his contemporary and fellow Viennese Sigmund Freud, Tandler was an assimilated Moravian Jew. He was born on February 16, 1869, the oldest of the seven children of Moritz and Rosalia Tandler. Probably under strained economic circumstances, the Tandler family moved to Vienna in about 1872. There they remained poor, moving often, and sometimes living in Vienna's poorest working-class districts. Tandler's father was sometimes without work; while in school, the young Julius worked as a bread boy to supplement the family's means. His life was dominated by an "unconditional striving to better himself" that was probably shaped by this early privation in a culture often defined by showy excess. Social movement was, however, possible in this setting in part because Bohemian and Moravian Jews were often given a German-based bilingual education, which, combined with the traditional Jewish regard for learning, encouraged assimilation into the mainstream culture of urban Vienna. But assimilation was never complete, and the working-class struggles that Tandler witnessed in his father's life and shared in his youth conditioned a dogged persistence that may have facilitated his affiliation with Austrian socialism.[5]

Tandler was not an outstanding student, but his best subjects were math and science. Perhaps for this reason in 1889, after a brief flirtation with drama, he enrolled in the University of Vienna's medical school. He was taken with the great Viennese anatomist Emil Zuckerkandl, who, in the university's hierarchical atmosphere, was said to befriend his students. After completing his medical studies, Tandler began research as Zuckerkandl's assistant in anatomy. When Zuckerkandl died suddenly in 1910, Tandler was offered his professorship and the leadership of Zuckerkandl's anatomical institute, a position that he held until 1934.

Biological science suited him, but Tandler was most comfortable with his science when it could be connected with reforms that would improve human health. Early on he steeped himself in clinical problems and the influence of scientific anatomy on medical reform.[6] At the time, the medical school at the University of Vienna was among the premier medical institutions in the world. Anatomy remained the dominant science in medicine, and Tandler's position as dean of the medical faculty gave him significant authority. His view of the relationship between medical science and reform was said to be widespread in

Austrian medicine at the turn of the century. Berta Szeps-Zuckerkandl, Emil's wife and the hostess to one of Vienna's most prominent salons—one where art, culture, science, and the medical establishment met—wrote: "Just as there is a pure Austrian culture that is completely distinct from the German, there is also a pure Austrian medicine."[7]

When Tandler was named director of the Peoples' Health Bureau (*Volksgesundheitsamt*) in the new Austrian republic in 1919, he had risen unusually far for a Jew, and his Jewish heritage was an issue. (See fig. 6.1.) Pan-Germanist sentiments that favored the union of Austria and Germany were strong after the war, and even converted Jews faced intense antisemitism. The Union of German Physicians, for example, protested Tandler's appointment on the grounds that they would have preferred the appointment of a native ethnic "German," not someone they deemed to be an unqualified Jew.[8] After the Social Democratic Party lost federal power, but retained the governance of Vienna, Tandler became municipal councilor for welfare (*Wohlfahrtspflege*) for the city. For almost fifteen years, from 1920 to 1934, his views dominated social welfare programs for its citizens, particularly those directed at the youth. His philosophy is summarized in his still oft-quoted statement, "He who builds palaces for children tears down prison walls."[9]

FIGURE 6.1. Julius Tandler walking with his friend Alma Mahler-Werfel and her daughter.

Reproduced with the kind permission of Marina Mahler.

In February 1934, Tandler was traveling in China, where he consulted extensively on hospital and health reform. Several colleagues in Vienna realized that during the long train ride from China to Moscow he would have received no word of the February 12th workers' revolution in Upper Austria, a revolt in which mass demonstrations were violently suppressed by the national government. With no information on the upheaval, Tandler naturally planned to return home. Friends and fellow physicians intercepted him at the Moscow train station to warn him that he could be in danger and convince him not to return. They feared he would be imprisoned. Leaving them with the impression that he would travel to Prague, he in fact came back to Vienna and was imprisoned. His reflections, given later in Moscow, recall an international physician's conference that was occurring in Vienna at the time. As many as fifty American physicians in attendance protested that they would boycott the conference and leave immediately if the government did not release Tandler at once. Chancellor Engelbert Dollfuß feigned ignorance and released him. Tandler left immediately for China and eventually returned to Moscow. He died there of a heart attack two years later, on August 25, 1936.[10]

Anatomy, Physiology, and Sexuality

Tandler was best known in the field of anatomy. He published several important works in clinical anatomy, and his handbook served for years as one the of the field's standards. Though he ranged widely, sexual anatomy became his central focus, beginning as early as 1899.[11] The two branches of biology that fed medicine—anatomy and physiology—had remained separate through much of the nineteenth century. But Tandler's approach demanded their union in a scientific medicine that was also linked to the functioning medical clinic.[12] He insisted that anatomy must be a dynamic science, not one that just passively measured fixed and dead tissue.[13] Knowledge of lifeless organ structure could not explain life's functions; anatomy would be useless as a foundation for medicine if it were not based in the dynamic processes of life. Properly linked to digestion, respiration, circulation, neural reactivity, and glandular function, anatomy was not about postmortem structure but about life itself. Tandler also stressed attention to organic diversity; comparative anatomy and comparative physiology should play key roles in medicine. This devotion to the union of anatomy, physiology, and biology shaped his early approach to reproductive anatomy. So, just after the turn of the century, he coupled his anatomical research on sexuality with what became pioneering work in reproductive endocrinology.

As Zuckerkandl's favorite student, Tandler was on the border of Berta's famous salon. There he would have met such medical luminaries as the pioneering surgeon Theodor Billroth and the psychiatrists Julius Wagner-Jauregg and Richard Freiherr von Krafft-Ebing.[14] Wagner-Jauregg's study of cretinism would have had an especially significant influence. Cretinism (congenital thyroid deficiency) produced arrested development and mental deficiency, and it was especially common in the Alps. Wagner-Jauregg believed that the thyroid was critical for the development of both brain and body, and in the 1890s he proposed that thyroid insufficiency caused cretinism, altering a person's entire constitution. Pervasive community-based intervention was the key to eliminating the problem, he claimed. To ensure such intervention, Wagner-Jauregg worked with teachers and doctors, spearheading a campaign with the Styrian government to provide the treatments needed to reverse the condition. Especially profound mental, skeletal, and genital abnormalities could sometimes be eliminated when treatments were begun in young children. For more than a decade he used regular doses of thyroid extracts and iodine to treat affected children living in Alpine villages in the area.

The Zuckerkandls sometimes accompanied Wagner-Jauregg on his Alpine trips, and Berta Zuckerkandl was astonished at the complete transformation she saw in some children after thyroid treatment. Their "monstrous heads," misshapen bodies, and "lifeless eyes," which, she said, had at first trigged revulsion, were altered "as if in a fairy tale where humans turned into animals by an evil fairy had been saved by a kind sorcerer."[15] These were the values of Austrian medicine at their best. Tandler began his constitutional research with Wagner-Jauregg, and he too, must have been impressed with both the amazing transformations in the children and the role of government in sustaining them. Several aspects of Wagner-Jauregg's work—the involvement of hormones, their effect on an individual's entire constitution, mental and physical, their sexual symptoms, and the impact of government involvement—presage important features of Tandler's career.

Beginning in 1909, Tandler published a series of papers on the effects of castration in mammals, which combined his interests in sexual anatomy, development, and physiology, and included important clinical work on humans. By 1915, he had become an acknowledged expert in gynecological pathology, reproductive physiology, and urology. The range of his work was great; it often seems to wander well beyond the field of anatomy. For example, much of the early work stressed embryology, which he taught from 1900. He completed work on freemartins as early as 1912, and he did comparative work on gecko brains, seals, deer, and moles that stressed evolutionary patterns.[16] His major hypothesis about secondary sexual attributes was an evolutionary one: he proposed that before being altered to distinguish the sexes, hormonally controlled differences in sexual characteristics had first evolved as general attributes of the species; only later in evolution did they come to distinguish the sexes.

Tandler's central focus, the biology of sexuality, emerged amid a complex tangle of cultural oppositions, some antisemitic and misogynistic, some Catholic and conservative, and some socially progressive, that shaped Viennese views on sexuality from 1890. Intellectual life in fin-de-siècle Vienna was imbued with tensions about the nature of sexuality. The art of Gustav Klimt and Egon Schiele, the literature of Arthur Schnitzler, and the psychiatries of Krafft-Ebing and, of course, Freud focused on unrestrained and/or repressed sexuality as one of the central issues of the age. One of Vienna's most disturbed sons, Otto Weininger, author of the antisemitic and misogynistic *Geschlecht und Charakter* (Sex and character), argued that women and Jews "had no souls." They had none of the transcendent autonomous self that was thought to empower human self-control, human reason, and will. Weininger was an unknown when he wrote, and he died immediately thereafter. But his elaboration of the nature of mind in women and Jews generated intense interest among Viennese intellectuals. And the attention given his ideas reveals the extent of the city's sexual tensions.[17] Zionism, the women question, debates about homosexuality, the sexual stereotypes of Jewish men—all helped condition the interest that Weininger generated. Historian David Luft, for example, notes that for Austrian men feminism was troublesome, not because it represented a serious threat to men's positions in society. Rather, men's preoccupations were with the power of female sexuality to threaten male rationality and produce confusion and fear about their own sexual identities. The result was vivid sexual expression in literature, art, and psychiatry that increased the intensity of the threat. Such widespread interest in sexuality was a defining feature of this richly intellectual, yet deeply divided society. Tandler's science—the anatomy and physiology of sexuality—has to be understood in this cultural context. He and Steinach were dealing with a new approach to the biology underlying one of the profound cultural tensions of their time.[18]

Tandler and Development

Tandler's diverse interests were not a collection of tangents reflecting the unrelated wanderings of a curious mind. Rather, the seemingly disparate threads were strongly related. Tandler thought deeply and wrote often about their interconnections. And, although there are a number of features that deserve more careful historical analysis, four important conceptual threads connect them. Understanding the interconnections helps parse the comprehensive and multilevel frames by which Tandler applied biology and medicine to society.

One is a fascination with development reflected in his emphasis on embryology and evolution. These were not sciences that could be separated from anatomy; rather, in understanding normal and pathological anatomy, developmental origins in dynamic patterns of change or in "errors" of development were keys to the etiology and causation of adult form and function (or dysfunction). This historical approach led him to incorporate comparative, evolutionary, and zoological work into his medical perspective. He understood anatomy and physiology in terms of the principles of evolutionary development that gave rise to individual and to species diversity in structure. So clinically it mattered that humans were born humans and not moles.

Second, Tandler's work regularly referred to and distinguished among several different levels of analysis. He discussed species, sex, race, and, most pointedly when addressing clinical problems, individuals. And he often coupled his reference to different levels of analysis with clear distinctions about which concepts applied best to which levels and with concern about the problems of moving among them. He stressed the complexity of integrating the diverse contexts of individual lives with broad scientific abstractions aimed at a more general—often a statistical—level.

Third, he regularly used these many levels to link developmental variation to the effects on anatomy of alterations in physiological mechanisms. For sexual anatomy, that meant endocrine physiology. Given his view on the relationship between anatomy and physiology, it is not surprising that new findings in reproductive endocrinology became central to his treatment of sexual anatomy.

Finally, although he stressed physiological mechanisms, especially in endocrinology, his work was not reductive. It embraced a holism that is best reflected in his pioneering work in constitutional medicine. It is here that his advocacy of the inheritance of acquired characteristics is clearest. All four aspects of his thought are well illustrated in an article written just as his political career blossomed, "Über Wachstumsstörungen" (On growth disorders).[19] The paper illustrates his care in distinguishing inferences that are appropriate to distinct levels of analysis, his dynamic conceptualization of development in relation to endocrinology, and the relevance of hormones to individual anatomy and constitution.

Tandler had already promoted reforms in medical education that placed medicine on a firm biological footing. Here too, he stressed that physicians must anchor their concept of growth in a "medical biological sense." But as a clinician, he was keenly aware of the distinct goals and explanatory styles of the scientific versus the clinical enterprise. He stated that scientific knowledge of anatomy and physiology can represent only a general sketch (*schematisierte Kenntnis*) for physicians, because it presents an idealized "representative of the species homo." Physicians must deal, not with group norms or ideals but with individuals whose unique histories require that doctors be prepared for huge deviations from the general principles given in physiology and anatomy. Tandler presented a concern for individuality as the physician's highest priority. "Individual variation, however, is above all that with which practicing doctors must be engaged, thus really the deviations from that which we have learned as the norm."[20] All levels were relevant; but doctors must deal first and foremost with individuals, even when their patients violated the law-like regularities accepted by medical science.

Tandler defined organic growth or development as all those processes through which "a living being achieves the proportions of individual maturity within a normal range of variation that maintains the functional ability to adjust to demands (*Beanspruchungsfähigkeit*)." Unlike inorganic accretion, organic development was not a process of addition. It always involved a "lawful readjustment of dimensions" occurring in an individual.[21] The adult is never just a large multiplication of the developing being, but a "novel creation (*Neuschaffen*) of specific dimensions." Growth was, therefore, synonymous with development, which always involved differentiation and reorganization. Moreover, the laws of development are infinitely complex, producing "complications, which are almost infinitely increased through circumstance," and which must take place within certain time requirements. Complexity and context require a dynamic approach, because, he said, growth is "a matter of rebuilding in the midst of operation."[22]

Using several examples, Tandler stressed the involvement of the endocrine system in the multilevel restructuring that occurs during development. "The growth of the individual is determined by his belonging to species, to a race, influenced by his affiliation with a family, constitutionally established, dependent, inside certain boundaries, by environmental influences, and in many cases regulated by the harmonious combined effects of the glands of inner secretion." The complexities abounded. Species, race, sex, and population effects could all be altered by individual uniqueness that was related to the environment and to the action of the glands. The developmental influence of the endocrine system even included interactions among the different glands within the system.[23] In the most complex cases, like cretinism, all the glands could be involved.

Tandler's work in reproductive endocrinology culminated in the publication of an award-winning monograph on the influence of inner secretions on sexuality, *Die biologische Grundlagen der sekundären Geschlechtscharaktere* (The biological foundations of the secondary sexual characteristics).[24] The work became an early and much-cited classic in reproductive endocrinology. It linked the double-gonad view and the secretory function of interstitial cells to his research on sexual anatomy. Nearly a decade after its publication, Tandler's emphasis on the potential for widespread developmental change as a result of hormone action was still apparent in "On growth disorders." The endocrine apparatus was central for alterations of development that he believed involved a wholesale restructuring of the body. "Of much greater meaning are those growth disturbances which concern entire organ systems and with them literally the entire body." In them, changes might be initiated in any of a number of systems, "whereupon the effect of the disruption is localized in a spatially distant effector. The last process is characteristic of growth disturbances that result from the disharmony of the inner secretory apparatus."[25]

Here too, however, he repeatedly stressed variety and the need to address each case individually. Even with involvement of the same glands, the specific nature of a disturbance and its long-term effect would vary from individual to individual. Tandler had studied children that were castrated at an early age (eunuchoidism). In them he saw, for example, that some children showed only a short-lived delay in development, while others remained in the eunuchoid state for their entire lives. The latter cases often involved persistent disruptions that required explanation beyond just the glandular disturbance in development. In discussing relationships among the levels, Tandler was clear that such individual variety often occurred within norms defined for the species and race. But unusual circumstances could produce individual extremes that altered these statistical categories. Further, the accumulation of individual extremes was, of course, critical in determining the average dimensions of the population, group, or race.[26] So individual variety fed analysis of the group just as much as group boundaries helped account for (but not explain) the individual. In discussing the variety of causes of re-formation involved in the development of dwarfs, for example, Tandler noted that although all involved developmental reconfigurations of bodily proportions, there were many distinctions among different categories of dwarfism that differed depending on the timing of the changes involved, the specificity of the system affected, and the combination of mechanisms abstracted for race or species with influences unique to an individual's life. Observations of the form and timing of re-formation were the keys to understanding both the laws of growth and development and the variations one saw as individuals deviated from those norms.[27] This convergence of influences and Tandler's attention to individual deviations made him sensitive to the vagaries of individuality and context. Poignantly, he stated: "Those belonging, for example, to the population of dwarfs are true dwarfs. At the same time, they are humans normally proportioned just in a diminished version."[28]

Constitutionalism and the Inheritance of Acquired Characteristics

Constitutionalism was a body of theory in medicine and biology that grew rapidly at the end of the nineteenth century. It was characterized by a holistic approach to the person that challenged the reductionism of the germ theory of disease and instead emphasized the interdependence among distinct features of an individual's body, things such as muscle tone, disposition to disease, and stature. The proponents of constitutional medicine assumed that somatic attributes of the body could predict not just physical demeanor but also disease propensity and a person's temperament and personality. In many forms of constitutionalism, these holistic amalgams of body and mind reflected an entire "habitus," the fundamental body type and character of a person.[29] Tandler was an early convert to the importance of constitutionalism's holistic and individualistic approach. He believed that the external appearance of a person (the habitus) was a central diagnostic tool for the clinician. And much of his work on sexual anatomy linked outer appearance to the functional effects of the secretions produced by the endocrine glands, especially the gonads. Writing before Steinach had completed his transplants, Tandler argued from his work and from numerous clinical findings that the formative and re-formative developmental effects on an individual's constitution were due to secretions, not exclusively to neural mechanisms.[30] Their effects began early in embryonic life, and they could be traced throughout development. Tandler's integration of an individual's constitutional habitus with clinical phenomena was part of a pioneering new approach to constitutional medicine that connected clinical anatomy and normal development to the endocrine system. Its secretions shaped individual attributes—the physical and psychological dimensions of the person.

Tandler founded and for over two decades edited the *Zeitschrift für angewandte Anatomie und Konstitutionslehre* (Journal of applied anatomy and constitution theory), which first appeared in 1913. And like his other themes, arguably more so, his focus on constitution integrates the seemingly separate themes that appear in his work. Different levels of analysis, the stress on development, and Tandler's insistence on tight linkages between anatomy and physiology, between structure and function, can all be situated within his emphasis on a comprehensive constitutional approach to the person. Importantly, that attempt also reflects the major role that the inheritance of acquired characteristics played in Tandler's early thinking and in his understanding of the concept of race. And his linkage of individual constitution and race, a connection that he understood in terms of the inheritance of acquired characteristics, framed his approach to the biology of race hygiene and human improvement.

The inaugural issue of the journal included an article by Tandler clarifying the concept of constitution, normal and pathological, and its relation to "condition."[31] German constitutional pioneer Friedrich Martius had argued that constitutional thinking was needed in medicine to offset the huge emphasis on microbes and explain why disease agents like bacteria caused disease in some individuals, but not others. Martius also addressed debates about causation in constitutional disease, stressing physical models of the internal energy of the body that could be affected by disposition and environmental influences, like nutrition, to alter disease susceptibility.[32] Paying proper homage to Martius's pioneering role, Tandler extended his framework to "normal" as well as pathological constitutions. He stressed the importance of portraying the individual as an integrated whole in health as well as in disease. And in normal constitutions, zoological and evolutionary themes were also needed to understand individual variability. So for Tandler, as for many constitutional theorists, the person was the core constitutional concept; but, though the concept of constitution should be applied only to individuals, understanding the person involved several levels, including sex, race, and context. Tandler's proposal for how they interconnected to form an integrated whole shows that, at the peak of his scientific career, he viewed the relationships among them in terms of the formative power of the environment, which he linked to the action of inner secretions.

Tandler's paper was a theoretical piece. In it he reconceived the relation between constitution and condition and related them causally in ways mediated by the actions of inner secretions. Constitutional attributes were those personal hereditary attributes that were left after species and racial attributes had been discounted. Most important, therefore, one could not take the collective attributes of a race or species and extend them from the group to the individual. To do so would be to confuse group with individual levels of analysis. Further, individual constitution was determined at the time of conception, and, in Tandler's view, it remained unchangeable within an individual's lifetime: different chromosomal combinations produced different individual constitutions. But Tandler introduced flexibility by way of his distinction between constitution and condition. Condition was also a concept that applied to individuals; it referred to the flexible attributes that were wrought by circumstance. An individual's condition could change and be changed by the environment. "What is in many cases termed acquired constitution cannot be subsumed under the term constitution. That which can be changed in an individual as a result of environmental influence is never his constitution, but his condition."[33] But from the moment of a person's existence (for Tandler, the moment of fertilization) and throughout his or her life, the two influences were combined. Individual habitus was shaped by hereditary constitution and by the variety of conditions to which an individual was exposed.

Having reformulated the relationship between constitution and condition, Tandler turned to the issue of race. Conceptual confusion and the misuse of words was "the order of the day" with the race question, and he stressed how difficult it was to distinguish racial from constitutional attributes. He argued for arriving at the greatest possible demarcation between them. This was because "a large portion of what we characterize as racial variation turns out to be constitutional variation."[34] Racial variation was more limited than constitutional variation. But the constitutional characteristics of a group could, even in humankind, sometimes be generalized as a racial attribute. Tandler paid close attention to how the transition from one to the other took place. Noting that he was taking a position on a topic outside his area of expertise, he stated that the proposed connection forced consideration of the "dangerous ground" of the inheritance of acquired characteristics. He asked: "Can a conditional attribute, therefore an acquired one, become fixed as a constitutional attribute, therefore a hereditary one? Further, in what way can a conditional attribute become a constitutional one?"[35] His own answer was 'yes' to the former question; but he contrasted his position with the alternative view that assumed no possible transition from condition to constitution—no inheritance of acquired characteristics. If one accepted his distinction between constitution and condition, which most would, but denied the inheritance of acquired characteristics, then habitus, demeanor, and temperament must be determined developmentally only through variation in the combinations of chromosomes dictating individual propensities and constitutional attributes. This, he said, was an approach that viewed "constitutional anomalies" as correctable only by an "unmediated constitutional hygiene" that would exclude those affected from reproduction, that is, by negative eugenics.[36]

This was only two years after Hans Przibram, Kammerer's boss, had asserted that the process of somatic induction, understood in terms of the gonad, had become a foundational question of biology. Przibram did not mention the double gonad—the anatomical separation of sexuality and reproduction. In 1913, however, there was no one better positioned than Tandler to interpret somatic induction in terms of the physiology of the double gonad. And he did. Admitting that he readily accepted the inheritance of acquired characteristics, Tandler continued his analysis by asserting that somatic induction was mediated by the secretory function of the gonad. Its inner secretions were critical for producing the influence of the environment on germ, and they were, therefore, also the means by which acquired changes in the individuals of a group—condition—could eventually alter the heredity of the group—constitution. Their powerful effects on outer demeanor were already clear in constitutional medicine, where somatic patterns were known to change drastically with alterations in circulating hormones. "It is, however, in my opinion very plausible that the inner secretory elements also transform the way to somatic induction, most especially the inner secretory elements of the gonads which we juxtapose with the generative element."[37]

The last phrase—"the inner secretory elements . . . which we juxtapose with the generative element"—is crucial. With it, Tandler was reiterating his position on the interstitial cells, a position that placed him with the French, who first proposed the hypothesis, and with Steinach, whose research was just appearing. Tandler's own work on the double gonad had begun in 1906, shortly after the French introduced the concept. In that year, Tandler and Siegfried Grosz received a stipend of 1,000 Crowns from the Austrian Academy of Sciences for investigations "on the existence and meaning of the interstitial substance of the sex glands."[38] Between 1906 and the publication in 1913 of their monograph on the development of secondary sexual characteristics, Tandler became closely associated with the secretory function of the interstitial cells. And by the time his article on constitution and condition appeared, he was the Viennese authority linking the cells to somatic induction.

Tandler cited Semon to introduce the question of somatic induction, but then brushed aside Semon's complicated theory. It was enough that it stressed physiology. Semon's summary of the evidence and, above all, the experimental findings of Kammerer, Przibram, and others, had suggested that induction was a process by which "somatic excitations were conveyed to the germ-plasm." In his view, endocrinology could fill in the details of that process. Tandler cited clinical evidence (overactivity of the pituitary or adrenals could, for example, destroy the generative portion of the gonad) as well as his zoological work on the testes of moles (which showed how fluctuations in the secretory part of the gland affected the generative, see fig. 2.1) to confirm the power of the secretions. Summarizing the relationship between constitution and condition, he wrote, "we are well justified in assuming on these and similar grounds that acquired, therefore, conditional attributes are transformed into inherited, therefore, constitutional [attributes] via the mediation of the inner secretory portion of the gonad."[39] The earliest Austrian emphasis on the double gonad and somatic induction was his, and as the new journal appeared, he stressed the interstitial cells as the key to conditionally shaped, but hereditary constitutions.

Tandler's paper on constitution and race hygiene began as a lecture presented to the German Society for Racial Hygiene in Munich in March 1913. The "special difficulties" that were raised by the concepts of race and constitution probably rested with this group's reception of his argument linking condition and constitution via the inheritance of acquired characteristics. Tandler had been invited to present the paper by Max von Gruber, the Munich race hygienist who chaired the German Society from its founding in 1910 until 1922. Gruber probably knew Tandler from the former's years as a professor of hygiene in Vienna. But the Munich group, which included the rigid hereditarians Fritz Lenz and Ernst Rüdin, was moving toward the interwar center of a radical German eugenics that based human progress on rigid Nordic racial ideals.[40]

We cannot be sure how the society responded to Tandler's analysis of race and constitution, but his attempt was a clear challenge to their direction. He admitted that he belonged to the camp accepting the inheritance of acquired characteristics, implying that many in the audience did not, and this changed how he looked at race. He then spelled out the implications for the practice of race hygiene. In the effect of individual condition on racial attributes also lay, Tandler said, the relevance of constitution and condition for race hygiene. Their interconnection inverted the preferred Munich method for achieving race hygiene. "If, as we have shown, condition in parents can become the constitution of the children, then condition hygiene is certainly indirectly the constitution hygiene of the children."[41] Because individuals defined groups like races, condition hygiene as proto-constitution was a form of race hygiene, and a very effective one. Via the inheritance of acquired characteristics mediated by the interstitial cells, he said, "condition hygiene" could be even more effective than traditional race hygiene. Tandler was defending the cells as the key to how constitution and condition informed the methods by which public hygiene might transform beneficial environmental outcomes into the future of the race. Because that which had been conditional could, in children, become constitutional, and because racial characteristics were defined by the average constitutions of the individuals who make up a group, children were a means to effect racial change. Even if an adult's constitution were fixed, altering conditions could alter a child's to produce lasting human improvement. "If condition in individual constitution is transformable into racial constitution, then a new and meaningful branch of social and medical endeavors is inaugurated in race hygiene," one that acknowledges the changeability of racial constitution and improves the conditions of individuals "to improve the character of a race."[42] Progress need not depend on selection among fixed racial attributes, as his Munich audience assumed, but on flexible development and environmental circumstances that, given the right conditioning, could alter racial constitution.

Tandler was describing the scientific foundation of his future social programs, programs that stressed improved health and living conditions for children. Hygiene might, he wrote, also strive to eliminate genetic damage by using a strictly constitutional approach (*unmittelbar Konstitutionshygiene*), which he did not reject; but he closed by noting that because condition hygiene did not depend upon excluding some individuals from reproduction, it had the advantage of satisfying those who objected to a selectionist race hygiene on the grounds that it violated individual rights and personal freedom. He concluded: "Hygiene as condition hygiene, however, is and remains the most powerful factor for the improvement in the individual constitution of the next generation and for this reason also of the racial constitution."[43]

These two works by Tandler established several important connections that predate their emphases by Kammerer and Steinach. Tandler stressed the environment in his distinction between constitution and condition. He then linked the relation between constitution and condition to somatic induction in the inheritance of acquired characteristics, understood in terms of the secretions of the interstitial cells. Having already accepted the idea that the gonad comprised two functionally distinct glands, he then stressed that via condition, the secretory portion could have far-reaching effects on a person's demeanor. And because its secretions had a special influence over the germ cells, it could also mediate environmentally based changes in the hereditary elements. As Kammerer would later argue in experimental and zoological rather than in medical terms, the interstitial cells could bridge constitution and condition, transforming heredity and enabling conditioning to shape constitution.[44] And by stressing that individual averages formed racial groups, Tandler too, linked the inheritance of acquired characteristics to "race" in a way that made race flexible. Condition hygiene was a basis for race hygiene; but this race hygiene stressed the changeability of racial characteristics made possible through the endocrine mediation of altered individual conditions. In the context in which the talk was given, the "dangerous ground" of the inheritance of acquired characteristics was not just the result of ongoing scientific debates about its validity in heredity. In Munich, it was dangerous because it rejected the rigid racial boundaries on which the Munich version of racial hygiene increasingly depended. It is noteworthy that in 1913, when he was a prominent scientist, but not yet a politician, Tandler took up the challenge.

Tandler and Kammerer may have met in 1907, when both worked with Rudolf Goldscheid to establish a unit for "social biology and eugenics" in the Sociological Society of Vienna.[45] But at the time Tandler was beginning to peak, and Kammerer was just beginning, so they probably had very little interaction. There was little personal ground for any interaction. The disciplinary foundations of their approaches were quite different, as were their scientific styles. Kammerer's was utopian and idealistic; he lacked caution and often overextended his science in the service of his vision of the ethical issues raised by selection and the undervaluing of workers. Tandler's style was, by contrast, pragmatic, anchored without emotion in objective science and its demonstrated effectiveness. His was less an idealistic integration of ethics and science than a set of practical conclusions based on his interpretation of the available evidence and the clinical and reform outcomes made possible by it. Even in his reference to individual rights in the 1913 paper, Tandler took no personal stand. If he agreed with the ethics of those challenging Munich-style race hygiene, he did not express his preference. Instead, he based his conclusion less on the fusion of a scientific and moral vision and more on "objective" interpretations of evidence that had great practical relevance.

In retrospect, this preference often left Tandler appearing "Janus-faced,"[46] sometimes criticizing the infirm and "inferior," while sometimes stressing flexibility and developmentally oriented social reform to defend children. How stable were these views? What motivated his vacillations, and how much did his approach change in the mid 1920s, as the weight of scientific evidence—or at least the conclusions drawn from it by some scientists—shifted more and more toward selectionist methods of human improvement based in rigid genetics?

Political Ascent and Scientific Interest

Tandler's political star began to rise when he met with Emperor Karl in 1916. At that meeting, Tandler convinced the new emperor that the war had produced such horrible problems that it would be necessary to establish a division of the government to administer programs to limit the spread of disease, build housing for the homeless, and provide nourishment and child care for the citizens. The meeting signals a shift in Tandler's career from being primarily an involved scientist concerned about public health to becoming an influential municipal politician with the responsibility to implement it. That shift did not, however, take Tandler out of active involvement with science. As his political career blossomed in the 1920s, Tandler maintained close contact with his science, including the science underlying the interstitial cell question.

By the late teens, Steinach was the acknowledged Central European expert on the secretory function of the interstitial cells. In 1917, Tandler sent Steinach a letter in which he asked about a figure from one of Steinach's most important publications, the one reporting his successful creation of experimental hermaphrodites. Steinach had grafted both male and female gonads onto the same muscle layers in castrated infant male guinea pigs. The figure showed microscopic sections of the "puberty glands" of these animals. Alongside atrophied sperm tubes, male interstitial cells had increased and become embedded in islands of "lutein cell-like" female interstitial cells.[47] The gland, like the animals, had become male and female. Tandler asked if he might inspect the original preparations on which the figure was based.[48] His interest was still sufficiently great that he apparently wanted to see firsthand that the two cell types could develop in the same animal.

Scientists in Tandler's institute continued to address the problem of sexual development and the interstitial cells into the 1920s. In 1921, Marianne Stein and Edmund Herrmann reported the results of experiments that tested the impact of hormone injections taken from extracts of female corpus luteum on the development of the testes, the prostate, and the seminal vesicles in male mammals. Injections that began in infancy produced varying degrees of inhibition in the development of the sperm-generating cells, sometimes producing nearly complete degeneration and sperm death in adults. To varying degrees, however, as sperm production degenerated, both young and sexually mature males showed large increases in interstitial cells. This confirmed both their control over the generative cells and the effectiveness of female tissue on male development.[49]

Stein and Herrmann's report then addressed the controversial bisexual character of the interstitial cells (see Chapter 2). They examined a structure called the "Uterus masculinus," a dormant uterus present in male guinea pigs. Many investigators, they wrote, considered the Uterus masculinus a vestige of the Müllerian ducts, female organs that are present in embryonic form in males, too, but that in adult males become just atrophied rudiments of the embryonic tissue. Stein and Herrmann rejected the vestige interpretation of the uterine tissue; they had measured the effect on the male uterus of injections from the female corpus luteum. The tissue was very responsive. The extracts produced extraordinary growth of the organ, rendering it full of secretions. This "growth impulse" triggered the "capability" of the organ's secretory structures and, by implication, its function. Stein and Herrmann stressed the "lively appearance of secretory activity" in the organ rendered active by the female interstitial cells.[50] Their work and its emphasis indicate that Tandler's laboratory was still immersed in questions about masculinity and femininity that challenged existing norms of sexuality and could be considered consistent with socialist values. In the early 1920s, Tandler continued to pursue the interstitial cell question and to accept the validity of the idea that he had helped to pioneer.

Shifting Views of Hormones and Genes

Tandler's influence must be understood in the context of the hope for human improvement through the application of biology to society that was especially strong in postwar socialist Vienna. That hope had grown before the war in the mutually supportive terrains of Austrian positivism and the particularly Austrian brand of socialism called Austromarxism. After the war, conditions had deteriorated so that promise became necessity, and Tandler and his fellow socialists seized the moment to promote its implementation. Its centerpiece was the Viennese social experiment known as Red Vienna. Beginning after World War I in the early years of the Austrian First Republic, the union of developmental science and social policy that produced Red Vienna generated some of the West's most progressive working-class social reforms.[51]

The socialists' goal was to finally realize the promise of the Enlightenment: human improvement through the just application of science and reason to society. It was an old ideal that had been tried before, but with little success. This time was, however, different. The Austrian Social Democrats had withstood the splintering brought by the war's end, and as the Republic formed, they were "the strongest Social Democratic party in the world."[52] Their power in Vienna made it possible that, given the huge changes in biology in the late nineteenth and early twentieth centuries, profound improvement could be realized in a moment of great need. The advances were real; and the promise was enthralling, especially given the hardship, disability, and disease brought by war. With mature biology based in evolutionary thinking, the germ theory of disease, and modern knowledge of medicine, physiology, and embryology, the project could succeed.

For Kammerer, Tandler, and Steinach, the key feature of these scientific changes was the integration of development, endocrinology, and heredity implemented on a scale barely attempted before. Scientifically, their success depended in part on the promise that Jennings had glimpsed, but that he had abandoned after Kammerer's death: the validity of the inheritance of acquired characteristics. The concept was fully compatible with the political goals of the Viennese Social Democrats. It was imbedded in the early writings of major Austromarxist theoreticians, and by the mid-1920s, Tandler was near the political center of its relevance for the postwar Austromarxist promise. He knew and had greatly contributed to the kind of science that promised change; that was the point of viewing heredity as a flexible result of the diffuse action of the endocrine system. But just as his power solidified, the concept that could make his reforms and the Austromarxist promise possible—the inheritance of acquired characteristics—had become one of the most controversial scientific issues of the day.

By the early 1930s, both the science of heredity and the field of endocrinology had changed, and those changes affected the integration of science and politics that was at the basis of Tandler's (and Kammerer's) effort. In 1928, the Austrian League for People's Regeneration and Heredity (Österreichische Bund für Volksaufartung und Erbkunde) was founded to help the country avoid degeneration and promote improved mental and physical constitutions. The group was an explicitly eugenic organization, but one expressly designed without a distinct political character. Its founders made sure that all sides of the political spectrum were reflected in the wide range of views represented on its scientific advisory committee. Tandler was a member. But amid the political turmoil of the late twenties, as mass worker demonstrations and violence broke out, Tandler's party became less effective. It even feared its own constituency, as dissenting workers favored mass demonstrations and activism above what seemed like empty party politics.[53] As political tensions heightened, the league became less active. In 1934, it was transformed into the Union for Human Heredity and Endocrinology (Verein für menschliche Erblichkeitslehre und Endokrinologie); psychiatrist Wagner-Jauregg, physiologist Arnold Durig, and jurist Felix Tietze were its leaders.[54]

The new group illustrates a shift in scientific thinking about the relationship between hormones and genes—one that lacked the stress on flexibility that Kammerer, Steinach, and Tandler sought to build—and the rising career of Julius Bauer shows the tensions involved in that shift. Bauer too, was an endocrinologist; like Tandler, he linked clinical work to the reforms of an endocrine-based constitutional medicine. He rose to prominence in the teens; in 1926, he was named associate professor of internal medicine; and by 1931, he directed the first medical department of the Vienna Polyclinic. Tandler had been a part of his rise. In April 1918, Tandler received a letter from Martius recommending that Tandler include Bauer as an editor of the *Zeitschrift für angewandte Anatomie und Konstitutionslehre.*[55] Tandler accepted the recommendation, and Tandler and Bauer worked together on the journal until 1934.

Historian Veronika Hofer argues that the constitution concept served as a "boundary object" that bridged clinical interests with public policy, psychiatry, eugenics, and genetics, and Bauer illustrates how effectively these boundaries worked. In the third edition of his major work, *Die konstitutionelle Disposition zu innere Krankeiten* (The constitutional disposition to internal disease), Bauer stressed the multiplicity of causal factors combining both inner and outer sources that were involved in constitutional disease. Bauer, like Tandler, associated pathological "constitution" with a hereditary disposition to contract a disease. But Bauer's constitutionalism assumed a stronger genetic determinism, an emphasis that that served him well in 1938, as he fled the Nazis to eventually help establish genetic analyses of constitutional disease in the United States.[56]

Bauer represents a younger generation of medical science, one connected to Tandler's perspective more by a common interest in the idea of constitution than by an overlapping approach. As he worked to incorporate changes in the rapidly moving science of heredity into constitutional pathology, he criticized the neglect of genetics that he saw in discussions of the clinical effects of hormones. This emphasis would grow throughout his career. But in Vienna in the late 1920s, it was new, cutting-edge, and dressed in the unfolding discoveries of the new genetics. Speaking of an individual's constitution, Bauer wrote, "The psychophysical oneness and wholeness is evident in the fertilized egg and represented by the chromosomal apparatus as the primary carrier of the genes."[57] Heredity was an essential blueprint of the whole. Though Tandler and Bauer both valued the holism brought by constitutional medicine and its utility in diagnosis, their views of the relationship between hormones and genes are embedded in a different figure and ground. Tandler saw hereditary inflexibility and used hormones to add flexibility. Bauer saw unconstrained hormonal flexibility and stressed greater emphasis on genetic inflexibility.

Bauer was taking a modernizing step, but his advocacy ignored the formative developmental flexibility that, for Tandler, linked constitution and condition via the inheritance of acquired characteristics. In the 1924 edition of his work on constitutional dispositions to disease, Bauer acknowledged the centrality of debates on the inheritance of acquired characteristics for the constitution concept, but he did not seem to know what to do with them. Citing Tandler, Kammerer, and Steinach, he summarized the relevance of somatic induction, acknowledging that the possibility was not incompatible with Mendel's laws. But he could draw no conclusion; he agreed with geneticist Richard Demoll that opinions and evidence were so mixed that the answer was not "yes" or "no," but "yes" *and* "no."[58] He seemed to want to avoid the problem, perhaps because it complicated the use of heredity in diagnosis, perhaps simply because over the years increasingly more of his generation came to side with Weismannism. As Bauer aligned himself more strongly with genetic structure as an essential plan and less and less with the formative aspects of condition by which individual adaptation could become hereditary, the issue disappeared. With a terse brush-off he dissociated himself from the flexibility that Tandler had won using endocrine-based somatic induction to link constitution and condition.[59]

Bauer illustrates the transition that young scientists faced amid the huge shifts in the sciences of heredity during the interwar period. Well into the 1920s, German-speaking genetics hotly debated the existence of extra-chromosomal influences that qualified a pure Weismannism. Victor Jollos's work on *Dauermodification* (persistent transformation) is a case in point. As late as the mid-1930s, Jollos had evidence that adaptive resistance initiated by environmental exposure to arsenic in *Paramecium* might be inherited via cytoplasmic, not nuclear, mechanisms of heredity. Though Jollos's interpretation was not Lamarckian, others who still favored the idea seized on it as further experimental evidence for the possibilities that had been promoted by Detto, Kammerer, and others.[60] But questions of how stable such changes were and whether they could be replicated, fed criticism from *Drosophila* geneticists, particularly in the United States. Jollos's work effectively stopped when he was fired in the Nazi's "Aryanization" program in 1933. He left for the United States, where American preferences for the new genetics made it difficult for his views to gain prominence and support.

Any middle ground in the sciences of heredity between Kammerer and Weismann was complicated in the years between 1910 and 1934 by broad scientific uncertainty on many counts. Debates about adaptation in evolution, about the role of the cytoplasm in heredity, about the application of experimental work and artificial conditions to nature, and about mechanism versus holism were inconclusive. New disciplinary boundaries were forming in science that complicated the resolution of such problems, as different communities increasingly spoke different languages and dealt in distinct empirical paradigms. Clinical endocrinology became more separate from chemical attempts to synthesize hormones, and both distanced themselves from zoology. *Paramecium* and *Drosophila* genetics became separate from human genetics. The boundaries forged by specialization made what had been interesting zoological and evolutionary integrations based in comparative analyses irrelevant to the new micro-paradigms that drove specialization. In this context, well-intentioned scientific debates and boundary formation left much open ground for the infusion of contentious social issues.

Authority and the Seeds of Change

Tandler's implementation of reforms that were clearly anchored in the inheritance of acquired characteristics in 1913 overlapped with this period of enormous scientific and social change. Where was Julius Tandler in all this? As a politician, he spearheaded reforms that were intended to be eugenic, but that depended on social and evolutionary changes linked by the connection between endocrinology and the inheritance of acquired characteristics. How did his thinking change as he stepped into politics? He gave us an early view in 1920. As Tandler took on the national responsibility for the enormous task of restoring the health of the new county, he republished his 1913 paper "Constitution and Race Hygiene." The 1920 version was much shorter, but most of the content of the paper is unchanged. Both the flexibility brought by the inheritance of acquired characteristics and the role of endocrine hormones, particularly those of the gonads, are stated in nearly identical words. And the key process of somatic induction was still stressed as the means to ensure adaptations that reduced race hygiene to condition hygiene. But Tandler was no longer just an influential academic physician. He was the head of the national government agency in charge of health and welfare. And this version of the paper was published in the Reports of the National Peoples' Health Bureau (Volksgesundheitsamt). Presented in this venue by a man in his position, his stress on the flexibility of heredity across generations and the superiority of condition hygiene amounted to a mandate for the biological foundation of his reform policies.[61]

Tandler was keenly aware of the need to encourage the people, build hope, and convince them of the sincerity of their government. He strongly believed that effective reform could not be left to private individuals or agencies. It was the state's duty to fulfill its responsibility to the people, and effective scientific intervention must reside with the state. But the people must trust that their government has their well-being at heart. Many of Tandler's writings and appearances, indeed his entire public image, were framed to accomplish this. And the early 1920s would have been the perfect time to take advantage of the passionate advocacy of a man like Kammerer, for whom the well-being of the workers was primary. But there is no evidence that Tandler ever answered Kammerer's letter. More puzzling is the fact that in his many writings advocating his approach for administrators and the public, Tandler never referred to Paul Kammerer's writings on endocrinology and productive eugenics. The idea was based in their shared appreciation for the power of endocrinology, the theory of the double gonad, and most important, the validity of the inheritance of acquired characteristics, which Kammerer had strengthened with the experimental evidence of the heat rats. Without it, the relationship between condition and constitution could not be a justification for Tandler's condition hygiene. Yet the phrase "productive eugenics" never entered his discourse.

Around 1920, when the reissued and shortened paper appeared, there may have been simple reasons for this. Tandler was a very busy man, and Kammerer had only just published the results of the heat rat collaboration with Steinach. But by 1926, Kammerer's views on the problem had repeatedly appeared in his many books and articles on the topic. They were widely discussed, and Tandler could hardly have missed them. But much else had changed by the late 1920s, as the world economic crisis once again aggravated the conditions of life in the recovering capital city. Selectionist eugenics implemented by the removal of the unfit from the breeding population was by then endorsed by the Supreme Court of the United States—the country of individual liberty. And selectionist eugenics had emerged in Austria, some of it linked to the conservative and racist approach gaining ground in Germany. And just as political and economic tensions in socialist Vienna worsened, Kammerer's suicide in 1926 dealt a serious blow to the credibility of the key concept—the inheritance of acquired characteristics.

As his reforms were implemented amid great political controversy, Tandler might have benefited from a surrogate mouthpiece whose voice was anchored in his policies and in Steinach's strong evidence on the interstitial cells. Yet Tandler was silent. And his silence raises important questions about how his approach to reform changed after 1926. How did he respond to the kinds of changes that Bauer advocated, linking endocrinology to an inflexible, but more modern concept of heredity? When, as a prominent politician, he was faced with enormous political tensions that demanded compromise, did Tandler turn to a more straightforwardly selectionist approach to human improvement? If not, why did he avoid aligning his solution with Kammerer's productive eugenics? Scientifically, it was his eugenics too.

7

Tandler's Eugenic Enigmas

Julius Tandler's eugenic beliefs remain a series of enigmas; historical treatments of his reforms run the gamut. Some see him as a socialist hero, friend and savior to the working class, a physician of the people who worked tirelessly to improve the lives of working-class children. Others condemn him as a eugenicist, wielding political power to "reform" and remove what he saw as damaging and "inferior" people. Doris Byer states, "The decisive problem remained: who determined the border between the 'natural' and 'social' environment? The question of the boundary is the question of power." Her treatment, however, acknowledges only natural selection as the perceived basis of evolutionary change, and she combines perceptions of hereditary progress with views of evolution that are linked solely to genetic predeterminism. This was certainly the case for many eugenicists; but, as Kammerer's productive eugenics shows, not for all. For Kammerer and Tandler, the border between the social and natural was entirely fuzzy, and quite different messages could be taken from the power of biology, depending on how that line was drawn. This is not to deny that Tandler and others wielded great power over people's lives, power they sometimes applied impersonally and with the potential to manipulate. Rather, it presents a more historically nuanced view of the motives and the biology behind that power.[1]

Gerhard Baader and Monika Löscher present Tandler from a more measured middle ground. Baader acknowledges both the Janus-faced character of Tandler's language, which sometimes linked his ideas to negative eugenics, and the foundation of his reforms in the environmental stance of his science. Tandler often referred to race hygiene and to selection; but unlike much German race hygiene, he also relied heavily on the inheritance of acquired characteristics. Baader presents Tandler's reliance on use inheritance as a key difference between postwar Austrian and postwar German eugenics. But he interprets that difference using the distinction between negative and positive eugenics. In Baader's view, Tandler was a positive eugenicist whose reforms promoted the greater breeding of the fit. A more contextualized view of Tandler instead acknowledges the extent to which he was, or was not, employing a version of Kammerer's productive eugenics. Tandler, too, believed in the productive power of development. He had even anticipated Kammerer in framing that power via the interstitial cells and somatic induction. But by the late 1920s, acceptance of the inheritance of acquired characteristics was in decline, and the political situation surrounding that decline was dangerously complex. Löscher's work emphasizes the wide range of views represented by Austrian eugenicists, including the powerful Catholic Church. She places Tandler with the socialists and notes his reluctance to use force and his emphasis on voluntary means of control. But she also stresses the ambiguities in Tandler's language and the potential for abuse implied in his social engineering. She captures those ambiguities succinctly: "One can always refute Tandler with Tandler."[2]

I agree; Tandler's language confuses our interpretation of his goals. But I also argue that few accounts of Tandler fully explore the impact of debates about the inheritance of acquired characteristics and their relationship to the ethics of responsibility in his thinking. Some authors seem to assume that because he recognized the dangers of "poor breeding," Tandler must therefore have sought to increase negative selection to weed out the unfit or improve positive selection to promote the fit. But at least in the early postwar years, selection of any kind was not Tandler's preferred method. Recognizing the dangers of unregulated selection did not entail the use of selection to reverse the trend. As discussed in the last chapter, though he considered them, Tandler usually rejected them, preferring his condition hygiene.

In what follows, I link Tandler's science with features of his reform philosophy to interpret these ambiguities and situate his view of nature and nurture in the context of the political tensions and eugenic debates occurring among the scientists in interwar Austria. The theories of sociologist Rudolf Goldscheid, on whom both Kammerer and Tandler relied, provide an important key. Tandler drew on Goldscheid's Darwinian sociology and Kammerer's extension of endocrinology to promote somatic induction. He regularly referred to the former, but almost never to the latter, despite numerous similarities between his and Kammerer's views. Examining the enigmatic relationship between Tandler's task and Kammerer's productive eugenics reveals the political tightrope on which Tandler had to balance. Political compromise required that he reach diverse audiences; yet he had to project an aura of scientific authority that would engender broad public support at a desperate time in the city's history. Both the opportunity for political compromise and the scientific foundation underlying the success of his reforms were in flux. At some level, Tandler probably knew this. But he could not afford to appear uncertain; so his rhetoric was crafted to be vague. In this context, public association with the controversial Kammerer could have been risky, whether Tandler accepted Kammerer's ideas or not. Successfully negotiating this complex sociopolitical terrain required a balancing act that led to apparent reversals and confusion.

The mid-1920s also saw a deepening of political tensions among the socialists and between the socialists and their most prominent opponents. I compare Tandler's justifications for his reforms in the early versus later years, the Tandler of First Republic optimism, as Kammerer's endocrine solution was being developed, with the Tandler enmeshed in the political chaos that ensued after revisions of the socialist program in 1926. The new party program endorsed force, if needed, to maintain order among the workers. Indeed, violence entered the scene with the spontaneous workers' revolt of July 1927. Eighty-nine lives were lost and several hundred people were injured as protesting mobs of workers burned the federal Palace of Justice in Vienna. Historians interpret this event as a turning point. It symbolized one of the Social Democratic party's greatest fears—that it could not control the enraged mobs of the workers it hoped to represent.[3]

These events occurred just as the widespread discrediting of Kammerer's work began. It must have been a difficult time for a tepidly political, reform-minded biologist, dedicated to the health of the young and pragmatically bent on success in reform. Was Tandler's environmental optimism challenged? Did his rhetoric change when Kammerer's scientific reputation was compromised? Did the political turmoil and the concurrent changes in science, which increasingly challenged the inheritance of acquired characteristics, alter Tandler's view of how to implement the reforms needed to protect the working class? If not, why did he never acknowledge the overlap between his programs and Kammerer's outspokenly socialist and antiracial stance?

Power and Desperation in Postwar Vienna

The Emperor Franz-Joseph had ruled the Habsburg Empire from 1848 until his death in 1916. His reign saw enormous changes in what, in 1867, became a dual monarchy, the Austro-Hungarian Empire. The new alliance required the dissolution of the German Confederation, and, though formal separation from Germany meant a significant loss of influence, a new constitution was adopted enacting many liberal reforms. Eventually, however, the reforms of the 1860s and new political crises of the 1880s gave way to a confusing mix of factions, which, by the turn of the century, included the rise of ethnic nationalism among the empire's diverse ethnicities, the emergence of democratic socialism, the invention of political antisemitism, and the creation of Zionism.[4] The result was an extremely volatile political situation: on the eve of World War I a diverse multinational state entered a war based in nationalism, the magnitude and effects of which were entirely unanticipated on all sides. During the war, the eighty-six-year-old Franz-Joseph died, the prime minister was assassinated, and a new emperor was crowned just before the empire fell. Austria-Hungary was a mess.

In Vienna by 1907, the liberals had long since retreated and three political parties dominated: the Social Democrats, the Christian Socials, and the more and more marginalized pan-Germanist Party. The charismatic mayor of Vienna, Karl Lueger, had founded the Christian Social Party just before the turn of the century. Conservative, officially antisemitic, and only loosely associated with the Catholic Church, the Christian Socials were a municipal party that appealed to the values of family and self-determination, while at the same time promoting numerous services through significant technological modernism. Holding a majority in the city by 1900, the Christian Socials literally electrified Vienna, bringing many modern conveniences to its citizens. They also actively sought to turn their constituents, members of the artisan middle and lower-middle classes, against the Jews—Lueger is considered the founder of political antisemitism. The party regularly portrayed the proletariat using nineteenth-century stereotypes of men who were parasitic do-nothings, content to let taxes levied on the middle class pay their way. The strategy worked well. The rise of Lueger's Christian Socials at the time when Austromarxist socialism was being elaborated largely by Jews defined what historian John Boyer calls the "great chasm" in Viennese politics before the rise of Hitler.[5]

Tandler was building his medical career and was not very politically engaged until he became dean of the medical faculty in 1915. But even by 1905, the seeds of tension that he would later face already divided the Social Democrats and the Christian Socials. In 1907, political complexities in the city and in the Austrian half of the empire were magnified when the imperial Parliament enacted universal manhood suffrage. Voting had been restricted to wealthy landowners and middle-class wage earners. With universal male suffrage, the franchise was extended to all men who had lived in the state for one year. The reform was promoted by the Social Democrats, and Lueger, who was mayor from 1897 until 1910, supported it against the conservative wing of his own party. Though in the process he demonized both the workers and the Jewish intellectuals supporting them, the change established the Christian Socials as an enduring federal and a municipal power. Their new strength outside of Vienna enabled them to unite with the more conservative Catholic parties that dominated the rural Alpine provinces in a quasi-united front against the socialist power in Vienna.[6]

Though it was not waged in the city, the First World War had an enormous impact on Vienna and on its relation to the new Austrian Republic. The Treaty of Saint-Germain, which ended the war between Austria and the Allies, stripped the former empire of most of its crown lands. The Czech Republic, Slovakia, the South Tyrol, and parts of the Balkan states all became independent or were ceded to others by the terms of the treaty. As a result, the war disrupted administrative structures that had been the main source of order, and neither of the two major parties anticipated the social, economic, and political changes that resulted. Far from its former status as a large multinational, multilingual empire, "Austria" now consisted of only 23 percent of the area of the former Cisleithania (the Austrian half of the dual monarchy). The effect was psychologically devastating. One-third of the empire's German-speaking citizens were suddenly living on foreign soil. Vienna, once the shining imperial capital, had to endure, in historian Helmut Gruber's words, the "psychological shocks to the metropolitan ego caused by the city's diminished status as a capital of alpine yokels."[7] Its resources and much of its industrial base were gone. The rail system was badly disrupted by altered national borders, and food and coal were in desperately short supply. Many of Austria's former provinces, new nations like Czechoslovakia, as if in retaliation refused to trade with Austria. Even most Austrians considered their new little country *lebensunfähig,* not able to survive on its own. "The Austrians had no pride in being Austrians and took little interest in their small country."[8]

Political theory often stressed the impact of economic differences, and they were real. But the tensions between the parties cut deeply into cultural values that affected even the most politically uninformed. Arguably the two most pervasive cultural flash points were religion—especially the role of the Church in the schools—and nationalism. In the multinational Habsburg Empire, ethnic and linguistic identities that coexisted with nationalist allegiances could be distinct from the state's official administrative structure. But at the war's end, when many of the crown lands became independent states, what became Austria had a primarily German cultural identity, and most political parties favored union with Germany, which was prohibited by the Allies. So the new Austria struggled with its identity. In this context, issues of religion, language, and nationality became inflamed, as the relationship between the ethnically diverse capital city was pitted against the rest of Austria's largely rural and more homogeneous citizenry.

The character of the competition between the Social Democrats and the Christian Socials also changed. Lueger died in 1910, and the war brought increased agitation for alliance with German nationalist parties. As war dragged on, conditions in Vienna were horrible. Disease and homelessness were rampant; street protests motivated by hunger, privation, high meat prices, and the cold sometimes led to violence. Soldiers, shell-shocked by war, took relief in the city; those in transit and war refugees from the east changed Vienna, as the city's population swelled to almost a third of all Austrians. Middle-class citizens and politicians on all sides feared disorder on the scale of the Bolshevik Revolution in Russia. And divisions among neighbors—shopkeepers and their customers—made Vienna itself a front, as hunger pitted farmer against consumer, neighbor against neighbor. The postwar result among the Christian Socials was a resurgence of political and popular antisemitism that was strengthened by closer ties to the Catholic Church. Ignaz Seipel, a Christian Social leader and eventual postwar prime minister, was a priest. By the end of the war, the polarities between city and country, between right and left, between middle class and worker, and between Catholic and Jew had deepened, as the capital city tried to forge a new relationship with its reduced and divided Austrian constituency.[9]

Parliamentary rule had been suspended in wartime; Emperor Karl restored it in 1917. But any promise of postwar imperial and legislative order was short-lived. In October 1918, following the defeat of the Austro-Hungarian army in Italy, parliamentary delegates from the traditional German-speaking parts of the empire formed a provisional national assembly and declared the "Independent German-Austrian State." The emperor abdicated as the war ended in defeat. A tenuous coalition between the conservative and antisemitic Christian Social Party and the Austromarxist and worker-oriented Social Democrats governed the new Republic of Austria. The government had burst forth from the ashes of the seven-hundred-year-old Habsburg dynasty at the end of a devastating war and with very little advance preparation.

In 1918, Tandler was installed as undersecretary of public health in the federal Ministry of Labor. The job did not last, because the two-party coalition of Christian Socials and Social Democrats did not last. It failed in just over a year. The Christian Socials joined with the resurging pan-Germanists, while the Social Democrats retreated to Vienna to run the city until they were driven out in 1934. Removed from federal power, Tandler became municipal councilor for welfare and social administration in Vienna. To understand his role, it is important to appreciate how much modern science shaped the Austromarxist view of social reform. Their worldview was based in a fundamental fusion of biology and the social sciences; theorist Max Adler described the tradition of Austromarxist socialism framing the Viennese social democracy as "scientific socialism."[10] Seipel, postwar leader of the Christian Socials, portrayed the Austromarxists as not just a political party but as "a community with a worldview (*Weltanschauungsgemeinschaft*)" parallel to that of the Roman Catholic Church.[11] Because of his foundation in biology, that worldview thoroughly informed Tandler's work; and in 1920, with an aura of authority brought by his strong scientific reputation, he stepped into these tensions to restore the health and welfare of the citizens of Vienna.

"Red Vienna"

Though the foundations of Austromarxism were laid before the war, Vienna in 1918 was largely a symbol of what had been. The statues crowning the magnificent architecture of the Ringstrasse, the music of Beethoven and Mozart, the benevolent monarch with his distant but fatherly love of the people, the empire itself—all were gone or profoundly challenged. Rather than a grand empire, Vienna now led a small and insignificant nation. But the Austromarxists brought new hope to the city, hope that they believed would transform Vienna.[12] They would apply rational scientific principles to shape a new social order for the creation of a *neue Menschen,* a new humanity, one not defined by the harsh economic divisions that had prevailed in the empire. In 1922, the Social Democrats formally separated the city from the province of Lower Austria and declared Vienna to be its own province. As a result, they controlled both the municipal and the provincial governments of Vienna and could levy the taxes necessary to implement the "socialist experiment" of Red Vienna. Workers would be shaped into a new humanity through the application of natural laws, often linked to the philosophy of physicist and physiologist Ernst Mach, which stressed evidence and rejected secular and clerical metaphysics.

Tandler, who often referred to the *neue Menschen,* was the most prominent scientist among the party's elite. As the man responsible for health and social welfare in a city devastated by disease and hunger, he was not just a central player implementing public health programs. Among a group of theoreticians who spoke often of science, but had never actually done any, he *was* the scientist. He was, therefore, also a symbol that embodied the scientific approach. To the public, if not to his colleagues, Tandler represented the promise of effective science in a moment of great need. In birthday tributes to Tandler, the main socialist newspaper the *Arbeiter-Zeitung* quoted the socialist leader Victor Adler: "When we won Tandler, the party made a splendid catch."[13] Amid intense fears for the future, he wielded his authority with a strong hand, and he must have been aware of the importance of guarding the scientific authority that underlay their hope for a better future.

Historian Helmut Gruber has called many of Tandler's reforms "municipal family supervision"; he rightly stresses the potential for abuse in the socialists' strongly top-down image of the worker family, an image dominated by middle-class patriarchy. Reformers aimed to "integrate the worker family into the norms of bourgeois life and thereby to assure stability and peace in the social order." But they sought that stability by exaggerating the decadence of worker life and imposing a model of the orderly, decent, and disciplined family that enshrined bourgeois values.[14] As a result, the socialists' intrusions into workers' private lives could be very far removed from working-class reality. For example, in defining the new worker woman in ways that sought to relieve her of the triple burden of work, motherhood, and household, the socialists recommended that women avoid purchasing jewels and instead buy sewing machines and vacuum cleaners. They missed the reality that working-class women could afford none of these. When the power structure deemed that children were at risk, the city could take them away from their families. Sometimes their measures were harsh; social workers would threaten mothers with removing their children if conditions did not improve, and the procedures used to remove a child sometimes began with early-morning surprise arrivals that seemed like police actions.[15]

Family problems usually implied issues about sexuality, and Gruber notes that there was a "predominant eugenic strain in virtually all discussions of the sexual question." Great stress was placed on eliminating what were perceived as sexual negatives (e.g., prostitution and homosexuality), and "attempts to lay the ghost of decadence and sexual waywardness within the context of social neglect" were rare, as the social origins of such problems were discussed only among a select group.[16] As Gruber asserts, this may have been true for many. But as I hope to show, it was not usually the case for Tandler, who regularly stressed the evils of poverty and social neglect. Many analyses of eugenic biopower in Austria do not take into account the varieties of eugenic models that were extant in the complicated political environment of the interwar period. And they rarely address the impact of changes in science—particularly those involving the inheritance of acquired characteristics—on the debates between the socialists and their opponents. For Tandler, Kammerer, Goldscheid, and others, neglect of social and economic conditions was the foundation of the problem, and remedying it depended on the biologically transformative effect of the environment.

Tandler's Reforms

Tandler was in charge of implementing reforms related to hospitals, disease, housing, education, and the family. His central focus became the well-being of children, seen and approached through what he called population politics or demography (*Bevölkerungspolitik*). He divided his commitment between the need to administer the war's wounded and fight diseases, like tuberculosis, syphilis, alcoholism, and malnutrition—all immediate postwar threats—and establishing the reforms that would enhance the quality of life for future generations. Tandler managed a large bureaucracy put in place to run many new services, such as the Marriage Advisory Bureau (*Eheberatungsstelle*) and the Mother's Advice Centers (*Mutterberatungsstelle*). Child Welfare Offices were installed in each of Vienna's twenty-one districts; they monitored nutrition, administered tests to detect syphilis and tuberculosis, and they vaccinated children. Reforms were aimed at the workers, where the financial need was presumably greatest.

But they were open to all, especially at war's end, when much of the city was starving. Tandler's division also coordinated a large increase in the number of kindergartens, which provided the education and the daily child care seen as especially valuable to enable both parents to work in times of desperate financial need.[17]

Many of the offices created to administer these reforms were populated by welfare workers, predominantly women, who had been trained for as much as two years in schools set up specifically to prepare them. A School Physician's Service was established to supervise regular examinations of children. The service kept records on mental and physical health that were sent to Vocational Guidance Offices, which then offered advice and testing on job suitability for worker families. Clinics were set up for dental hygiene. Physical exercise was encouraged through the Youth Welfare Department, which organized playgrounds, built municipal pools, and established recreational facilities to make time in the country possible for all children. Various offices were even given the guardianship of illegitimate children, who, Tandler specifically stipulated, also deserved care and education. All activities were centralized in a Children's Transfer Office (*Kinderübernahmsstelle*).

The task was enormous, and the centralized network was huge and (in Tandler's mind) critical. His solution was a model of progressive social engineering for which, by 1930, he had become internationally famous. Especially during the declining economic conditions in the early 1920s and the depression that followed, the beneficial outcomes were many. But, as Gruber and others have noted, the concept, the structures, and the widespread desperation also opened the door to abuse. The supervision of the Central Child Reception Office, for example, effectively eliminated the problem of poor children begging on the streets. But it did so by maintaining tight control over where children were housed and when or if they were allowed to return to their homes. Many were not. Financial rewards and gifts were given to those who agreed to receive tests or treatment, potentially permitting coercion among desperate people unwilling or unable to resist. But the administrative and personal desire on Tandler's part for individual growth was also genuine. For example, though he perceived alcoholism to be a very dangerous problem, educational efforts to treat it were set up to permit recovered patients to promote abstinence among those still in treatment.[18] So workers could help reform workers. If coercion were possible, in Tandler's mind at least, the effort would involve some measure of self-determination.

Class tensions and the political situation greatly complicated the rhetoric and the reality. Middle-class fears of the economically stressed working classes had begun to grow before the war. By 1900, Marx's concept of the *Lumpenproletariat,* revised by hereditarian views of crime and poverty in the lower classes, fueled middle-class fears of the "dangerous," "slothful," and "immoral" workers.[19] In Vienna, as disorder that had been triggered in working-class suburbs spilled across the Ringstrasse, political and social stereotypes were deepened when counter demonstrations by the right targeted both newly arrived Eastern Jews and assimilated Western Jews. In September 1911, orderly demonstrations of hungry workers organized by the Social Democrats ballooned out of control and several people were killed. The bourgeois press interpreted the riots as an adolescent revolt of ruffians (*Lumpenproletariat* and "flotsam").[20] The Christian Socials exploited these fears by depicting an out-of-control working class that was inflamed by a Jewish elite. The horrendous psychological, economic, and medical conditions that accompanied the transition to republican democracy after the war further aggravated an already tense situation.[21]

Maintaining order in a way that promoted the health and well-being of the people was, in this context, a monumental undertaking. The goal was clearly an interventionist social engineering that employed political power to restore and improve the people. Balancing the benefits that, for example, checked the dangerous spread of tuberculosis with minimizing the potential for abuse was a delicate task, at best. Tandler was a well-meaning man, convinced, perhaps too convinced, that he had a solution, and doggedly determined to carry it out. How else could the people be given hope? Viewed in the most positive light, Tandler's conundrum entailed the implementation of homogeneous, middle-class notions of health, prosperity, and morality in a heterogeneous society that was searching for much-needed stability amid the collapse of its main principle of order, the empire, and the shock and devastation of war.

Quality and Quantity in Population Politics

Julius Tandler almost never referred to Kammerer's work; and he never publicly endorsed Kammerer's interstitial cell hypothesis of somatic induction, a hypothesis which drew on his own pioneering research. However, several of his writings point to Tandler's acceptance of the logic of a productive eugenics based in the inheritance of acquired characteristics and an emphasis on instilling a communal sense of collective responsibility. Tandler also incorporated a notion of quality similar to that embraced by Kammerer and Goldscheid (see Chapter 5). Like Kammerer, Tandler stressed the nurturing of fewer, higher-quality young, not the relative selection of those inherently of high quality; and like Kammerer, he sought to instill biologically a collective ethic, which he extended from Goldscheid's sociology into biology.[22]

Tandler's early writings on social demography (*Bevölkerungspolitik*) laid out his views on the relationship between quality and quantity.[23] He distinguished historically between quantitative and qualitative Bevölkerungspolitik. The imperialistic quantitative Bevölkerungspolitik of Rome aimed for military power by building manpower. More recent commercial Bevolkerungspolitik was based in what Tandler saw as the flawed teachings of Thomas Malthus. It also sought more hands, but for factory work. A third form, "national Bevölkerungspolitik," like the ancient imperialistic form, advocated greater numbers for armies and defense. Each emphasized the quantity of people without regard for the attributes or the social configuration (*Aufbau*) of the people. Each, Tandler said, differed from socialist Bevölkerungspolitik, which would stress, not numbers but the quality of nurturing directed toward the young. If anything, "striving for numbers" played into the hands of the capitalists; it had no effect because so many children died before they could join the ranks of the proletariat.

The solution to human improvement was to improve the quality of living conditions, and this required altering the environment. Tandler described it as "making each individual productive and capable," a part of the struggle to fight capitalism.[24] Socialist societies could do this because they stressed the "humane existence of the individual, his cultural aptitude for the existence of a people." Improving the conditions of each individual's life would also, through the inheritance of acquired characteristics, eventually change the quality of humanity as a whole. The stress on an improved collective was certainly there; but achieving it depended on "constantly increasing encouragement in the nurturing conditions (*Aufzuchtsbedingungen*), the care of the young, [and] the intention to influence their general cultural adjustment."[25]

In his 1916 work on war and population, Tandler questioned the then popular idea that war was a selection factor that tested the ability of a people. Instead, war brought social conditions that produced biologically significant human damage. Because the military had converted schools into hospitals and aid stations, the youth had been neglected at just the time when their fathers were conscripted and their mothers were forced to do factory work. Because the children lacked domestic oversight, prostitution and youthful crime had increased. Other problems, such as declining nutrition, also jeopardized individual quality. Tandler argued that too much stress was being placed on quantitative breeding (*Quantitätszucht*), and what passed in many cases as qualitative breeding "often misses basic biological premises." Implying that it was not possible to advance quality by breeding, Tandler wrote, "we should at least try to promote quality in upbringing. That takes [a system of] child welfare according to *biological and social principles*, for which a range of sound measures are available." Ever sensitive to the advantages of timing in development, Tandler stressed that with children it was never too early to begin.[26]

Horrible postwar economic conditions—increasing costs of rearing a child and the stress on working mothers—had compromised the quality of family rearing. The solution was for the community to offset the parental burden and invest in the development of existing children: "We are instead today in a situation in which we should consider much less the birth of humans than the maintenance of the already born, because here it is a question of people who already exist, [and] who we must sustain."[27] There were certainly economic advantages of doing this—it increased organic capital and the return on investment in children, and Tandler regularly stressed them. But the war had so worsened economic conditions that many had been forced into poverty. And a major source of the war's impact had been the entry of women into the labor force. The war economy left working women with a double burden: earning a working wage and caring for the home and children. It was no wonder, Tandler wrote, that they collapsed under the weight of both and sometimes left their jobs. "Whoever knows the circumstances under which these women bear their children, and whoever knows the conditions under which these women carry out their child rearing, must stand in wonder at the heroism of the sex." Both men and women had the ability to manage these problems, but circumstances must encourage them. Tandler stressed several economic issues, including taxes, agrarian conditions, and increases in production. But it was Goldscheid's thinking that connected economic value to human value: "The leaders of the state must finally realize that the state is determined not only financially and legally, but also organically, that it is determined by its organic capital."[28] War had decreased the quality that parents were able to bring to rearing their young, jeopardizing Austria's organic capital. But, even with a declining birth rate, help could come from transformations in economic patterns enacted by the state.

Tandler cited two requirements needed for his solution to work. The first was cooperation among several scientific disciplines. Medicine would be the key piece in a more comprehensive and rational program. Medicine must, however, recognize the social, economic, ethical, and pedagogical dimensions of the problem. The interrelatedness of the biological with social conditions that were the provinces of sociology and economics meant that social workers would not just identify the needy and ill. Rather they were part of the solution; they would directly shape the abilities and moral sense of the needy. The second requirement was the importance of nurturing a sense of responsibility among those seeking help. Tandler wrote: "Responsibility is a requirement not only of those executing help, but also of those seeking it. Self-evident ethics (*ethische Selbstverständlichkeit*) is a proven help means test, if welfare is not to devolve into blind good deeds."[29] Without using his terms, Tandler therefore incorporated Kammerer's emphasis on the potential of reciprocal cooperation (*gegenseitige Hilfe*) and the resulting natural ethical drive it would produce. Responsibility was not just a middle-class obligation; the person being helped must also become responsible, with his or her sensibilities honed by example and pedagogy. Qualitative change had the advantage that it instilled in the individual a natural "consideration of a sense of responsibility for the next generation."[30]

As a result, people would eventually accept as a matter of course that self-improvement was linked to the nurturing of children. Altered economic conditions would feed individual responsibility, which in turn strengthened the ethical impulse of future generations.

The troubling ambiguity that regularly punctuated Tandler's language is apparent in several of these early works. He often referred to measures such as selection and breeding alongside quality child rearing. Children always came first; so in his mind, harsh measures that were applied to the adults of one generation were sometimes justified by their protective value for the children of the next. "Qualitative improvement must begin with the population, with children."[31] And his stress on child welfare was regularly framed with confusing references to mainstream eugenics. There were, he believed, some lost causes—some *Lumpenproletariat,* who were "bereft of any kind of sense of responsibility."[32] He also referred to the impossibility of raising the quality of war invalids and those unable to work. Postwar conditions defined a zero-sum game in which maximizing benefits to damaged or ill adults would drain resources from children. Tandler even assumed some fixed hereditary differences, such as those implied in his distinction between those segments of the proletariat with and without feelings of responsibility. His techniques would work with that portion of the working class—and he clearly stated that there were others—which "possessed feelings of responsibility and aspirations to quality (*Verantwortungsgefühl und Qualitätsehrgeiz*)."[33]

But, though his language was compromised, his conclusions and his recommendations were not. He did not advocate selection. Indeed, he regularly suggested that selection would not work. Flexibility was possible and preferred. His plan of implementation rested on strengthening individuals (*Ertüchtigung*) by altering economic circumstances that transformed the conditions of child rearing. Altering the economic conditions that influenced how people raised their children and shaping the responsibility they felt in doing so were economic and organic tasks: economic privation caused organic damage, and economic support instilled organic responsibility. The disastrous conditions brought by war could be reversed as the state instilled quality through the improved conditions of economic life (*Wirtschaftsleben*).

Although it has a less completely spelled out biological foundation, Tandler's program is only sensible eugenically if, beyond a focus on the positive selection of the fit, it expresses the spirit of Kammerer's productive eugenics bolstered, as for Kammerer, by Goldscheid's sociology. The developed quality of each individual contributes to progress by changing the quality of the species through flexible heredity and a potential for mutual aid. Quality in future generations is reshaped by environmental reform, not by relative changes in numbers. Tandler placed the reform burden on condition hygiene, noting the shortsightedness of quantitative approaches and stressing that the state must realize that the "acquisition and increase of this capital is surely connected to the organic."[34] "Capital" here referred not to monetary acquisition, but to organic acquisition that, wrought by condition, had become hereditary.

Reproductive Restraint

Tandler's constitutionalism (discussed in Chapter 6) shows the clear link that he drew between the inheritance of acquired characteristics and the importance of "condition hygiene." But by the late 1920s, much else had happened. In the works described above, Tandler wrote from the perspective of a scientist establishing a foundation for public health reform, while he was still a newcomer to politics. By 1926, however, he had become completely enmeshed in the tangle of Viennese and national politics. In that year, he tackled the complex legal debates that were raised by attempts to reform Paragraph 144, the section of the Austrian constitution that made abortion a punishable offence. In an address before the Union of Austrian Judges, Tandler treated abortion as a eugenic measure.[35] His presentation of its relation to other solutions gives us a window into his thinking as a more seasoned politician. Stressing attention to social "etiology" rather than to statistics, he presented a sympathetic portrayal of working-class mothers to examine abortion from the perspective of socialist Bevölkerungspolitik. He was a critic of abortion, but not of the women seeking it. The frequency of abortion among working-class mothers resulted, he said, not from a failure of their sense of responsibility but from the reverse. Mothers' keen feelings of responsibility for the children they already have lead them to consider abortion so they might better provide for a smaller family. Whoever has seen these women "can hardly be of the opinion that here frivolity (*Leichtsinn*) is the mainspring of abortion."[36]

Despite his positive assessment of women's motives, Tandler's view was that unregulated abortion was not biologically defensible because an individual's life begins at the moment of fertilization. For this reason, he rejected the many calls to repeal Paragraph 144. He instead imagined an ideal law, one that might grant legal abortion, not as a right of the mother but to save her life or for the good of society. The embryo, he reasoned, was a part of human society, not exclusively the possession of its mother. Society had the right to sacrifice the embryo, just as it sometimes might sacrifice soldiers on the battlefield, and just as an embryo must sometimes be sacrificed when its life is weighed against its mother's. Tandler identified three situations that would justify legal abortion under his ideal law: medical indications, eugenic indications, and social indications. Two points are notable here beyond simply Tandler's views on abortion. First, Tandler as good as rejected eugenic indications, which he equated with selection against "the inferior (*Minusvarianten*)." He maintained that "the eugenic indication plays at the moment practically no special role; it would however have high demographic (*bevölkerungspolitischen*) interest, if the scientific foundation of eugenics were better established." The implication of the last phrase is that this was not yet the case. Echoing the concerns raised by his historical treatment of Bevölkerungspolitik, he viewed the social and medical indications as still resting on firmer ground. To explain, Tandler referred to the promising but tenuous field of heredity, which had "few such solidly established premises with which, in individual cases, we could definitively base the logical conclusion to abort a pregnancy."[37]

Second—and this was Tandler's bottom line—with this social right to dispose (with legal abortion) comes also society's obligation to maintain life and shape a child in ways that are livable and humane. That obligation must attend to the presence of humane conditions. If society were to acknowledge both sides of its responsibility—to dispose and to nurture—it would then also see the wisdom of rejecting an unregulated right to abort in favor of laws that specify the circumstances justifying abortion. Tandler then laid out a plan by which abortion requests might be decided by a state commission. But he noted that his scheme could only work if society were ready to fulfill its nurturing obligation to future generations. His own preference and his view that society's obligation to maintain life, once established, would win out were clear when he confessed that in many cases this ideal system would effectively convince mothers to have their children. They would know that society's obligation to nurture its children would support them. It was a utopian stance that would make the punishments endorsed in Paragraph 144 superfluous: regulated abortion was not a right, but its use could be inculcated as a generative ethic similar to Kammerer's *Gesamtwille* and sustained through Goldscheid's advanced evolution (*Höherentwicklung*) (see Chapter 5).

There had been a time, Tandler said, when mothers were the sole nurturers of small children. But "mothering" was no longer the exclusive domain of women and the family; and the burden could not fall entirely on mothers. Greater economic demands on individuals required that society was obliged to help by supporting environmental reform at home, though clinics, education, and schools. Reformers, Tandler said, must appreciate the value of kindergartens, which provide relief from child rearing and increase the possibility that, economically, mothers might "save and maintain the family." That the general public had agreed to share the burden of nurture was reflected, he said, in the over one hundred working kindergartens in Vienna, which daily took care of over eight thousand children.[38]

These comments reveal that as more and more tumultuous events occurred, Tandler's views did not shift to a more radical hereditarian ground. He continued to rely on Goldscheid—"The goal and duty of demography is the management of organic capital, which is represented through living humanity bound in a commonwealth (*Gemeinwesen*)." These were not the actions of one single-mindedly dedicated to eliminating the Lumpenproletariat. Throughout, Tandler understood the quantitative and qualitative distinction in a developmental and "productive" sense. He recognized that although negative selection occurred in human beings, this did not entail the presumption that selectionist solutions were therefore the best way to proceed. Rather, development could promote progress, whether negative changes had been wrought by selection or by horrible conditions. Tandler's focus on development, like Kammerer's, emphasized the role of the environment in increasing the quality of individuals already born. Through them the inheritance of acquired characteristics would alter the people and the species. And, though he still decried "blind relief (*Wohltäterie*)" in which private organizations unscientifically supported the breeding of beggars, he stressed and implemented healthy environments for the mental and physical development of children. A society must have children to work with, so births must increase; but eugenics through kindergartens is a far cry from forced or even voluntary sterilization. Speaking of the children in those kindergartens, Tandler wrote, "whoever knows how vast are the completely life-influencing impressions directly bound to this age, will understand the importance of exerting this influence."[39]

Eugenic Competition

The establishment of a section for "social biology and eugenics" in the Sociological Society of Vienna tells us that Tandler was not alone; others shared his eugenic goals, many in a way that drew on Kammerer's biology.[40] Kammerer and Tandler joined the group to promote their vision of human improvement framed in environmental and evolutionary terms. And that vision was different from others that emerged in Vienna after the war. By the mid 1920s, a competing camp was promoting a different notion of eugenics that was well represented in academic medicine and anthropology.

Thomas Mayer has described this richly connected pan-Germanist wing of the eugenics movement in Austria. The group focused less on economics and quality rearing, and instead advocated a genetic determinist and racially based approach that relied on selective breeding. Its most prominent members were the anthropologist Otto Reche, who, from 1924 until 1927, headed the racially oriented Vienna Society for Race Cultivation (Wiener Gesellschaft für Rassenpflege); internist Heinrich Reichel, who often represented Austria at international eugenic congresses; and internist Rudolf Polland, who linked positive and negative eugenics to Nordic racial superiority. They were a German nationalist group whose strong genetic determinism aligned them with the racist approach developed by Rüdin, Lenz, and others in Munich and distinguished them from the Social Democratic eugenicists in Vienna. The contrast is clear in the German nationalists' early advocacy of forced sterilization and in their racially based antisemitism. Jews were not allowed in the Vienna Society for Race Cultivation.[41]

Mayer suggests that the German nationalist camp left neo-Lamarckians, like Tandler and Kammerer, behind. But he addresses neither the issue of somatic induction nor the opposition's regular challenges to the idea of the inheritance of acquired characteristics. Their writings show that they took pains to reject the notion, revealing the strong pull that it held for their competitors in power in the city and the importance they gave to challenging the threat that it posed. In his analysis of the role of racial health for the peoples' future, for example, Reche asserted that the germ cells were the only living beings that were maintained and carried over to the next generation. Only those attributes that were bound to the germ-plasm could be inherited. "This extraordinarily important recognition means that there can be *no* actual 'inheritance of acquired characteristics.'" One could imagine that individuals' attributes might influence their germ-plasm; but that did not imply their hereditary transmission. Reche reasoned that the main requirement for the healthy maintenance of the "stock" was maintaining the fittest germ-plasm. In his scheme, therefore, the idea of fitness was transferred from organism to germ. And if there were any doubt about Reche's selectionist meaning, he clarified it by illustrating with the hereditarian practices then in use in the United States. The United States, he said, showed how through legal forced sterilization the implementation of "race cultivation" might succeed: "Ambition with the help of race cultivation has made the American people first in the world."[42]

Reche also refuted Tandler's arguments on the relationship between constitution and condition. The fit and healthy human constitution must be saved, but it could not be saved by Tandler's condition hygiene. "Because *acquired attributes* are *not* inherited, all *individual hygiene,* all sport, all physical conditioning and education, is directly of advantage only for the *individual.*" Reche was Tandler's scientific peer in Vienna in the mid 1920s, and his references to sport, education, and conditioning are direct refutations of Tandler's view of how, via use inheritance, the development of individual quality could indeed influence the race and species. Reche shows that for those for whom rigid racial purity was central and for whom group attributes dictated individual attributes, the concept to refute was Tandler's condition hygiene, a concept that for both sides could be justified only by the validity of the inheritance of acquired characteristics. Reche's refutation appeared in print twice in identical form: once in 1926, when it was published after being given orally at the Vienna Society for Race Cultivation, and again in 1938, as the Nazis entered Austria.[43] On the earlier date, Reche was still in Vienna and the condition hygiene established under

Tandler's leadership was being actively implemented in Red Vienna. Reche did not name Tandler, but his pointed rejection of Tandler's views reveals the importance to the competition of challenging the approach being implemented in the city in 1926.[44]

Rudolf Polland approached the German nationalist eugenic program as a dermatologist/internist who was closely linked to constitutional medicine. In a 1924 periodical aimed at working physicians, he stressed the untreatability of constitutional problems, noting that recognizing this was especially important for German doctors. Efforts such as treating diabetics with insulin were a "false humanity," inconsistent with the physician's task of striving for improvement in the constitution of the next generation through "reproductive selection and selective breeding (*Fortpflanzungsauslese*)." Selection, the "unyielding regulator of all life . . . was unremittingly at work, unconcerned with the sorrow of the individual," and physicians must follow suit.[45]

Polland referred to Tandler's distinction between constitution and condition and, like Reche, rejected the inheritance of acquired characteristics. But Polland specifically mentioned somatic induction. Noting the difference between somatic and parallel induction and, importantly, the proposed connection between the former and the endocrine glands, he asserted that, like so-called mutations, somatic induction was unlikely to play a significant role in human beings. He instead stressed unchangeability: "The individual hereditary factors represent unchangeable units which, though they can be combined with one another in the most diverse ways, always however remain the same."[46] The clear implication was that no positive hereditary change that depended on environmentally based flexibility was possible in humans. By 1930, Polland had illegally joined the Nazi party in Austria.[47]

Interwar eugenics in Austria was a politically charged mix of diverse and competing views. But, as historian Paul Weindling has stressed, for a time it was also an arena in which "coalition and consensus" drove "regenerative social welfare." Tandler and his competitors in the German nationalist camp were "polar opposites," who in this shifting attempt at coalition tried to work together.[48] The Austrian Society for Public Health (Österreichische Gesellschaft für Volksgesundheit) reflects the attempt. Formed in 1927 to create an organization uniting diverse political interests around the goal of improving health, the society was an administrative organ for dealing with the deteriorating economic situation. The collaboration included representatives promoting Catholic and pan-Germanist agendas, but in Vienna its works reached the people largely through the public persona of Julius Tandler. The collaborative effort continued in the founding in 1928 of the Austrian League for Regeneration and Heredity (Österreichische Bund für Volksaufartung und Erbkunde), also a politically diverse union of right and left. Tandler and Reichel were both on its scientific advisory board. By the early 1930s, however, these attempts at collaboration were more rhetorical than real. A major subtext of their efforts was the impact of conservative and racist agendas. As they targeted the Jewish "race," one side increasingly threatened the existence of the other. Amid the chaotic political negotiations after 1927, Tandler's advocacy of a controversial scientific idea and his self-styled image as a symbol of hope made him a target for the other side. He had to deal with the competition but still convincingly promote optimistic, authoritative, and effective public appeals to health and reform. By the late 1920s, the growing appeal of racial and German national themes left him tottering on a slack rope between changes in the sciences of heredity and the growing potential for political disaster.

Gefahren der Minderwertigkeit: Tandler's Productive Eugenics

In the 1890s, British Darwinist Thomas Huxley had warned against the ethical violations and oversimplicity of the eugenic proposals then popular in England. By 1920, many considered Huxley's caution to be outdated in light of new advances in genetics.[49] But this was not the Viennese view. Those advances were still in debate, and Tandler too, expressed caution about whether "eugenic radicalism" sufficiently acknowledged the complexities of applying genetics to humanity. His caution reveals that he simultaneously answered to three masters. The first was the political demand that he accommodate conservative critics to maintain his political pragmatism and get something done; the second was the requirement for caution amid a great scientific debate—one that led him to maintain his position, but not eliminate alternatives before a scientific consensus had emerged. The third was his conscience, which above all valued medical action in the face of disease and misery. Especially under drastic conditions, his conscience required that he balance the benefit to individuals against the good of the people and, for him, that meant favoring children. Answering to such distinct and at times conflicting demands produced the dangerous language that he sometimes used and the reversals and ambiguities that may have been a conscious, if not terribly wise, strategy.[50]

Tandler's ambiguity was partly scientific. He accepted the conclusion that heredity was flexible. But many scientific questions remained unanswered. What were the limits of the flexibility associated with the inheritance of acquired characteristics? Why had Weismann's experiments removing the tails of mice not worked, while Kammerer's experiments shortening the siphons of sea squirts apparently had?[51] Kammerer himself acknowledged that not all persistent environmental adaptations in one generation were retained in the heredity of its descendents. But he had no answer to the question of why some did, but others did not. And by the late 1920s, the scientific tide was turning toward Weismann's view. That meant that Tandler's scientific foundation was weakening. As he tried to build the public's confidence in the possibility of improved health and a rehabilitated working class, how much detail could he provide on what were undecided scientific questions about the mechanisms of heredity? Would the uncertainty engendered by public awareness of unresolved scientific debates undermine the confidence and self-determination that he sought to instill? Tandler was surely not immune to the fears engendered by the mass demonstrations of 1927. But his discourse must be seen as partly calculated political compromise, partly rhetoric, and partly scientific uncertainty that acknowledged a growing international consensus within genetics that challenged his view of flexible heredity.

There is no better place to look for the changes these tensions may have engendered in Tandler's later views than the paper most often cited to reveal his eugenic sympathies: "Gefahren der Minderwertigkeit" (Dangers of inferiority). It is Tandler's clearest eugenic statement, and it, too, is fraught with ambiguity, an ambiguity increased by his desire to both shape evolution and improve the quality of existing individuals: "Qualitative Bevölkerungspolitik of course also postulates on selectionist grounds a given quantity of individuals, [it] however demands above all the qualitative improvement of the individual and of the whole in the interest of human dignity (*Menschenwürdigkeit*) and of culture."[52] He refers to selection throughout the work; but his references are framed by two important distinctions: the distinction between the task of eugenics, on one hand, and the most workable solutions needed to realize that task, on the other; and the distinction between the long-standing human zeal to improve humanity, often manifest in the impulsive use of outdated or overstated tactics, and the measured caution needed to build reforms on modern and established science. He was clear that for most of history, the desire to complete the task had outrun the quality of the science behind it, and this was still a major risk.

Tandler wrote: "Consider that just the normal and therefore responsible caring parents remain childless, while the irresponsible, the drunks, the feebleminded, the criminal, bring many children into the world; thus one realizes the entire magnitude of the danger." The hazards implied in his title were clearly there, at least in principle. And he seemed convinced of their cost: he cited, for example, the institutional expense of caring for so many mentally retarded children, a cost that was greater because it aggravated the poor conditions, bad schooling, and undernourishment that compromised the potential of normal children. "Human society is therefore justified, will and should it stand its ground, to impede the increase of the inferior." This is his assessment of the problem—selection might be justified—and from our modern perspective, it conveys a sense of Tandler's moral failure.[53]

But Tandler then wrote that, apart from the goal of eugenics, we must consider how best to carry it out. "The nearest way is to impede the breeding of the inferior." This he noted, citing U.S. laws forbidding some marriages, was not a new idea. And it was not a good one. His own study of these methods—his recognition of their inadequacy—he said, had led him to implement educational methods. The preferred means to achieve the task was not by force or any legal measure, but through institutions such as Red Vienna's Marriage Advice Bureaus that stressed education and acculturation. "To enact laws, above all of a social, even more of a eugenic nature, without the presence of a psychological inclination to their acceptance (*psychologische Aufnahmsbereitschaft*) in the people, means not only a failure of the legislative process, but also an impairment of the very [moral] tendency that lies at the foundation of the law."[54] Without measures that would instill and enhance individual responsibility, no law would work. But, once educated, those given advice would be transformed and meet their moral obligation to their progeny. He admitted that sterilization could work, and, because it had become medically much easier, it was likely to become more common. But, he regarded such radical eugenic methods as dangerous and ineffective. So, though he acknowledged the dangers of "inferiority," he also rejected state supported legal measures in favor of education. And in the process, he praised the often-stigmatized poor. Education, he noted, would have great success with the majority of people, who already have "no lack of a sense of responsibility" for their children. He even included "many inferiors" who, once educated, would be duly concerned about the future of their offspring. "Even here first of all education (*Aufklärung*) is necessary and not the law." His preference and his policies reflected not the power of one individual over another, or of the state over the individual. Rather, he advocated the inculcation of an inner sense of moral self-accountability, manifest as caring for offspring, present and future.[55]

But Tandler's strong developmental emphasis implied that some adults were too far-gone to be rehabilitated; and for them sterilization might be an option. He included among them habitual criminals, idiots, and the badly epileptic. Given limited resources, he believed that stress should be placed on enhancing the quality of the young, even if that meant sacrificing the reproductive potential of those beyond help. One of the subtleties that we must bring to reading Tandler is his appreciation for developmental timing: when in the lifespan of a person can an individual be rehabilitated? In his view, some could not be; and that changed how the zero sum game of allocating resources should be played. Stressing development was both his strength and his failure because it entailed a moral compromise. Intervention, while possible, might be too late; and for Tandler any compromise must favor investment in the young.

This was more than just a political nod to cooperation with the other side. In a letter to his good friend Joseph Friedjung written after Tandler was exiled to

Moscow, he privately acknowledged the dangers of inferiority: "Inborn inferiors (*Minusvariante*) exist and recognizing and acknowledging their existence is not pessimism, but seeing clearly what is." But as late as 1936 he also underlined his belief that the current hereditary understanding of deviance, disease, and inferiority undervalued biological flexibility: "As long as medicine, in its understanding of heredity questions, stops with supernumerary fingers interpreted as if the arrangement and association of the ganglion cells can be innately flawed, it will always plough the stony earth of unchangeable defectiveness (*Fehlerhaftigkeit*) and [just] dreadfully slog away."[56] From the perspective of his pragmatic realism there was no contradiction here. He recognized the potential costs to children of maintaining hopeless "inferiors;" but he hoped for something quite different, something far more pervasive that might define the future of what could be. That hope was based on reversing what he perceived to be scientific error in genetics coupled with a deep regard for flexibility in development.

The scientific error was the view that heredity was unchangeable. In "Gefahren," even after noting the wisdom of sterilizing those beyond help, he again warned against eugenic radicalism, by which he meant sterilization: "Admittedly here too [with sterilization], too much radicalism is damaging." After acknowledging that some people are defective, he nonetheless asked rhetorically, "Who are the unfit?" Stating that the phrase "hereditary burden" had become a dangerous buzzword, he waxed poetically as he wrote: "Where a fleck of scientific gold becomes minted with the worthless metal of public opinion as a slogan, there the coins lose their worth." Tandler was stating that there was something wrong with the science that underlay increasingly broad public support of inheritable hereditary burdens among the "unfit." "Eugenic radicalism, which is preached on many sides, damages eugenics; the adherence to unscientifically founded indications compromises it." These are revealing statements. They show how far, in his view, the human zeal to improve humanity *still* overstepped the science of heredity; and they reiterate his reliance on the inheritance of acquired characteristics as a means to alter both individuals and the species.[57]

Many historians see Tandler's advocacy of education as a form of positive selection. But Tandler also deemphasized positive eugenics and called for an approach that "raises the worth of a single individual."[58] Effective eugenics did not depend on breeding more "superior" people, but rather on shaping a "higher quality (*hochwertiger*) humanity." His solution differed from positive selection in its stress on permanently enhancing the moral self-determination of existing individuals and thereby of humanity. In the second part of the paper, Tandler described the better science behind how to do this; and in it he reasserted the value of environmental reform anchored in his distinction between constitution and condition and understood as a result of the inner secretions of the endocrine glands. Negative and positive selection through marriage advice, education, and sterilization were modest methods of selection that could serve eugenic ends. But in his mind, even when used voluntarily, they were not preferred. His description of the science underlying his preference shows that he mistrusted positive eugenics. To illustrate how difficult it would be to promote the "breeding of higher grade humans" he returned to condition hygiene, again stressing its dependence on the inheritance of acquired characteristics.[59]

As he discussed how the normal constitution was inherited, he identified the great question to be whether "through the preferential treatment of the carriers of hereditary qualities" those hereditary qualities can themselves be influenced—"whether change in the soma can have an effect on the generative elements of the carrier."[60] He summarized the views of those who rejected the idea and who assumed instead that a new being is given entirely in parental chromosomes. Rhetorically he asked, is all that is possible already given in preexisting hereditary combinations? Tandler reacted to the question with uncharacteristic passion; such a view was "fatalistic resignation, crippled and not creative (*nicht schöpferisch*)!"[61] His reaction underlines the persistence of his belief in the adaptive power of individual development for heredity and it reiterates Kammerer's criticism that selection alone could never do anything new; it could not be productive. Both the word *schöpferisch* and the passion could have been Kammerer's. Stressing changeability, Tandler reiterated that all hygienic reforms "are able to operate through condition; therefore all hygiene is in this sense condition hygiene." He added: "At the moment [in Red Vienna] condition hygiene is extended indiscriminately (*wahlos*) to all humanity, and given that in it, as in every human activity, humanitarianism and mercy must rule, also to the inferiors (*Minusvarianten*). Condition hygiene does not ask about worthy or unworthy lives."[62]

The views expressed here had not appeared so assertively in his writings since 1920; and except for the added emotion, they are unchanged. Without using his phrasing, the approach with which Tandler ended the paper is nearly identical to Kammerer's productive eugenics. Importantly, in reiterating condition hygiene as the way to improved constitutions, Tandler also referred obliquely to Kammerer's fate. His brief allusion confirms, I argue, the similarity of Kammerer and Tandler's views; it shows that Tandler himself recognized it. After again rejecting selection, Tandler grudgingly mentioned the midwife toad scandal. "I am convinced that the acquired characteristics of parents, therefore conditional attributes, become the inherited ones of children, therefore constitutional [attributes]. The proof of this cannot be supplied [only] by means of a mangled experiment (*des verstümmelnden Experimentes*), especially when it has been repeated a thousand times."[63] The phrase "a mangled experiment" is a veiled reference to the fraudulent toad specimens. Tandler said no more about the scandal surrounding the fraud. Nor did he use Kammerer's name or comment on his guilt. There was enough other evidence that one experiment did not matter. Instead, he reemphasized the role of the endocrine glands in making a conditional state of parents into a constitutional one of their child: that is, he reiterated the endocrine solution that the two men shared: "When therefore a change in condition occurs which influences the harmonious cooperation of the glands of inner secretion, . . . this influence also extends to the generative elements, to the germ-plasm." The double gland, harmonious organic cooperation, an altered human morality, and creative productivity wrought by the inheritance of acquired characteristics—it was almost all there. Tandler's emotion carried him to the end of the paper, where he again rejected selection: "In the deep shadows of resignation, we cannot leave the fate of our children to the combinatorial elements of the chromosomes alone."[64] His desire to save the concept even extended to removing the shadows cast on it by the midwife toad scandal.

Why Tandler Never Cited Kammerer

With such agreement in their views, why did Tandler never explicitly cite Kammerer's impassioned pleas for an alternative "productive" eugenics? His evasive reference at the end of a work arguing late in the game for the inheritance of acquired characteristics was as close as Tandler got, and even that carries the tone of aggravated distancing. To understand the omission, we must understand the enormous differences between the two men. They were very different personalities; and those personalities supported quite different styles of science and distinct accounts of science to the public. Though, as biologists, both men felt deeply their sense of responsibility to humanity, Tandler was a cautious and authoritarian pragmatist, while Kammerer was an impulsive and rebellious visionary. Kammerer sought a philosophical universalism anchored in monism and in justice that was buttressed by his involvement with Freemasonry and the ethical sensibilities instilled by Jodl. Biology—symbiosis and the inheritance of acquired characteristics—was a part of the oneness that ensured universal equality. Kammerer's monism, his acknowledgment of the political nature of all science, and his appreciation of the oneness of subject and object instilled a broad philosophical vision that encompassed his science. Driven by artistic sensibilities as well as by science, he had a post fin-de-siècle disrespect for scientific objectivity. His moral and philosophical position eagerly framed his science; and he built his political and philosophical foundation into his science as long as they did not fly too much in the face of the data.

This was heresy in the rational discourse of dispassionate, objective science, and Tandler could not accept it. He was of an earlier generation—more an empirical realist. His 1926 paper on the legal issues surrounding abortion highlights the difference. In it, he warned that the rhetoric of religious, ethical, and other aspects of a person's worldview could contaminate the scientific approach to Bevölkerungspolitik. He implored his readers to separate their subjectivity from scientific objectivity—to remove political and ethical biases from their science. In "Gefahren," he wrote that demography dealt with "the highest inconvenient truths that one unflinchingly must see to implement biological logic unencumbered by preconception, or by deep-rooted feelings."[65] For Tandler, the subjective and the objective were separate, their differences anchored in dry enlightened reason. Science was not cultural; it was evidence based. Tandler could carry scientific medicine from Austria to China, Russia, and the United States, implementing reforms developed elsewhere, because their foundation transcended culture. Politics and culture did not "contaminate" objective science. Even nature and mind were separate for Tandler. Nature thrived on regularity, mind on freedom and unstructured irregularity. To resolve them he spoke only of an eternal desolation that he chose not to reveal publicly.[66]

Their differences affected how each man tried to reconcile nature and nurture. Kammerer's fusion of politics and science and the oneness he summarized in his paper on dualism and heredity (see Chapter 5) left them completely intertwined; there was no "them" there; "they" were one. This was not the case for Tandler. For adults of a given generation, condition and constitution were separate. Condition brought change, but an individual's constitutional fate was unchangeable and determined at the moment of conception. Nature and nurture were distinct components shaping life; they did not fuse. Rather, their separate components contributed to evolution, as environment (condition and flexibility) in the individuals of one generation contributed to heredity (constitution) in the next. Two separate elements could both be part of the same process; but they were not one. And for some damaged adults of an older generation, their separation could take the form of rigid typologies.

This greater separation framed Tandler's references to inferiority. He found it difficult to explain their connection using the concepts of the new genetics, perhaps in part because he was caught in a real inconsistency. By 1929, he confessed to himself his pessimism about the possibility of clearly conveying the flexibility of constitution to medical experts. In his diary Tandler comments on a lecture on constitution, which he gave that year at the Mount Sinai Hospital in New York. In it he dealt "reasonably with the stupidities (*Dummheiten*) genotype—constitution, phenotype—condition." But he was not satisfied with the reaction he got. "Only a few will have understood. . . . That is yes the old evil, constitutional therefore unchangeable."[67]

Their differences in scientific styles and personality probably had a great impact on Tandler's reception of Kammerer's work. Kammerer was too speculative, too given to reaching beyond the data to wax passionately about quixotic possibilities. Any aversion produced by such differences would have been aggravated by political tensions that required Tandler to at least appear to concede some points to the other side. He could not afford to undermine either his scientific authority in the eyes of the public or the pragmatic quest for political compromise. This meant balancing his preference for creative shaping via condition with some concessions to principles of selection. The emotion that he finally expresses at the end of "Gefahren," where he refers to the "deep shadows of resignation" stands out as an uncharacteristic retreat from his caution. One might consider positive and negative selection; but, as this emotional moment shows, Tandler finally defended his assessment of what was most important: implementing them would be ineffective and perhaps disastrous. For a moment, Tandler lost his scientific cool. None of this minimizes the danger of his selectionist language and the implications of his references to *Lumpenproletariat* and *Minderwertigkeit;* rather, it interprets the motives behind his ambiguity, motives that became more complex after Kammerer's voice was silenced.

Kammerer's reputation was all but destroyed after 1926. The data for his endocrine hypothesis of somatic induction could not be decisive and were ignored (see Chapter 4). By 1928, when "Gefahren" was published, the scientific tide had moved further away from the inheritance of acquired characteristics. Aligning himself with Kammerer could have further undermined the acceptance of Tandler's programs, which by that time faced strong competition. It was risky enough to act on a scientific concept that many then denied; there was no need to stress the possibility that it might be fraudulent. In this context, Tandler's 1920 presentation of the 1913 "Constitution" paper becomes significant. In the 1920 version, Tandler retained his by then outdated summary of Semon's theory and included only a brief mention (through Semon) of Kammerer. Semon's views, never as extensive as Kammerer's, were simply no longer part of the debate. The omission may suggest that Tandler was avoiding Kammerer well before the scandal.[68]

In 1917, Kammerer wrote a second letter to Tandler. In it he requested that Tandler send him a photograph of himself to be used in a popular lecture in which Kammerer planned to describe some of Tandler's research. Tandler had not replied to Kammerer's first letter, and his response to the second is very revealing. He answered that he could not provide a photograph because he possessed none and because "I don't believe that the projection of an author's picture is appropriate to raise scientific interest in any question."[69] The first reason is almost certainly false. Tandler was an oft-photographed man in interwar Vienna. Official portraits and informal photographs of him commonly appeared in newspapers and public documents; he very likely had at least one to send. He refused because he was avoiding association with

Kammerer's public voice. His second reason suggests why. In the early 1920s, he may have depended on Kammerer's science to justify his acceptance of the inheritance of acquired characteristics. But Kammerer was vain; he regularly put his handsome face on his popular books, in his articles, and on the advertisements for his many public lectures. Tandler found this offensive. For Tandler, scientific ideas stood independently of those who presented them. Evidence and objectivity brought their own power, without requiring support from images of the person who offered them. Tandler didn't like Kammerer's style: the brash, vain zoologist had insufficient appreciation for restrained and disinterested scientific objectivity.

Equally important, for Tandler to explicitly associate himself with the controversial scientist, who before and after 1926 was so aligned with the concept, could have been politically reckless. Tandler had to give both scientific sides lip service in the name of balance and political compromise. Aligning himself with Kammerer's views could have risked the attempt at consensus that superficially united the factions of Austrian eugenics. In this light, the phrase "mangled experiment" may even reveal Tandler's annoyance. Tandler was defending what he saw as a wrongly discredited idea that had been supported by an inappropriate kind of advocacy: Kammerer's. It was the man (and his mangled experiment), not the concept that was at fault.

One element that was prominent in Kammerer's approach was missing in Tandler's later writings. This too may reveal political motives for Tandler's distancing of himself from Kammerer. Despite his early view of the significance of condition hygiene for racial flexibility (and his competitor's reliance on Nordic racial superiority), after 1920, Tandler never used the logic of productive eugenics to challenge the race concept. Historian Robert Wistrich provides a possible reason why. Wistrich argues that the Social Democrats in Vienna had their own antisemitic element; it was apparent in their inadequate response to the official antisemitism of the Christian Social Party after the war. Antisemitism was strong among rank-and-file workers at this time, and the Social Democrats did not want to be seen as the party that was devoted to protecting the Jews. To do so might alienate their constituency. Further, as race science grew, the inheritance of acquired characteristics was increasingly seen as a Jewish idea (see Chapter 8). To advocate the concept and side with Kammerer's views on race might have placed Tandler among that wing of the Social Democrats that appeared too Jewish. Prior to 1926, Tandler could let Kammerer deal with the race issue and maintain his preferred distance. So, just as it avoided risking scientific authority over a scientific issue around which much debate swirled, Tandler's ambiguous language also enabled him to avoid appearing too Jewish. Better to let the controversial Kammerer, who could and would say anything, do the dirty work for him.[70]

Especially after 1920, as he sought to implement the reforms of Red Vienna, Tandler's career was fraught with tensions. He was a man of action. He often said that talk wouldn't help; reform depended on action and deeds. Implementing any change in the complex political context of postwar Vienna was difficult. Negotiating the convergence of both scientific and political conflict complicated the huge task of restoring the population to health and prosperity. Western science was changing dramatically; debates about fundamentals, like the nature of genes and hormones (both hypothetical entities at the time), made it a moment of great scientific uncertainty. Tandler, the scientist, would not commit unequivocally to one view or another in a way that appeared premature, at least not until things became extreme. In the early years, he stated his position clearly, but as a preference in an open debate as much as a conclusion, and almost always conceding a bit to the other side. Prior to 1926, Kammerer was the central advocate of the inheritance of acquired characteristics, linking it to Goldscheid, to socialist reform, and to a different kind of eugenics. Tandler could, silently and reluctantly, rely on Kammerer without incurring the risk of publicly endorsing extreme or overly subjective versions of the idea. So he acknowledged selectionist approaches, while promoting his non-selectionist reforms and his preference for flexible heredity. His language walked a fine line between his own scientific beliefs and the ideas of those presenting eugenics and public hygiene through selectionist logic. But it was a rhetorical line that his programs never crossed. His ambiguous language may have seemed politically expedient as long as he controlled the kinds of policies actually being implemented.

But by 1928, maintaining that control became increasingly difficult, as the move to selectionist eugenics grew. Now with few other voices heard, Tandler returned to the logic presented in his constitutional work of 1913, and strongly endorsed the role of hormones and the secretory portion of the gonad in somatic induction. In doing so and in stressing the creative power that had justified his programs, he endorsed the spirit of Kammerer's productive eugenics, even though, as a cautious scientist and shrewd politician, he neither endorsed Kammerer's elaboration of the mechanism underlying it, nor the label "productive eugenics" that followed from it.

8

Working Jewish in Vienna

Between the fifteenth and eighteenth centuries, Jewish people were expelled from the city of Vienna several times. Catholic prejudice, superstition, and religious intolerance combined with economic upheaval and competition to ensure severe persecution in which Jews were, for example, restricted to a few kinds of work and placed under great personal constraints. There had been repeated calls to ban Jews completely from the city. As late as the 1750s, the Enlightenment empress Maria Theresa refused to speak to Jews. As a result, many Jews lived outside of Vienna, in regions of the empire such as Bohemia and Moravia. This changed in the 1780s, when Maria Theresa's heir, Emperor Joseph II, enacted several new reforms. These included the Patent of Toleration, which, for example, allowed Jews to attend public schools. People were no longer required to wear yellow badges marking them as Jews; they might live in areas throughout the city (not just in ghettos); and they might enter any academic field of work. There were still restrictions (they did not have equal rights and had to pay to reside in the city), but the changes were great. Historian Bruce Pauley writes that "Joseph made Austria the first county in the world to grant Jews naturalized citizenship and to consider them permanent residents; it was also the first to allow complete toleration in religious affairs."[1]

A foundation for religious tolerance had been laid. But Joseph died in 1790, and strong reactions to the French Revolution, the conquests of Napoleon, and the revolutions of 1848 eventually combined with the traditions of Austrian Catholicism to increase prejudice and intolerance. Nonetheless, the influence of Viennese Jews on their society slowly began to grow. Tolerance grew too, especially under the influence of the young emperor Franz-Joseph, who ascended to the throne in 1848. The new constitution of 1867 removed most laws discriminating against Jews, and Vienna's Jewish population grew enormously. As a result, at the turn of the twentieth century many assimilated Jews saw the Habsburg reign of the 1860s and 1870s as their golden age. Historian Marsha Rozenblit argues that the presence of so many ethic, national, and linguistic identities in a single multinational state enabled Viennese Jews to become culturally German, politically Austrian in a non-nationalist sense, and to retain their sense of Jewishness as one of the many ethnicities present in the city and in the empire. Before the 1880s, the multinational and multiethnic diversity of the Habsburg Empire in effect made it easier to be Jewish in Austria-Hungary. But the triumphs were short lived, as deeply embedded stereotypes persisted and eventually even strengthened. By 1900, a surge in European nationalism joined with the rise of race science and the ideal of Nordic racial superiority to produce a reversal that facilitated the welcoming of Hitler in 1938. The monarchy itself had retained what Steven Beller calls "structural antisemitism," an administrative infrastructure that, far from granting universal rights, restricted Jews from the "symbolic heart" of the state: the Habsburg bureaucratic club. The result reinforced a persistent sense of Jewish difference.[2]

Indeed, a new logic and a new political demographic gave modern antisemitism newfound virulence. Wilhelm Marr coined the term "anti-Semitism" in 1879. His argument was racial; it juxtaposed Jews to Christians on racial, not religious grounds. Jews were, he assumed, biologically distinct from "pure" Germans, and no amount of assimilation or conversion could change that. Even baptized Jews with little religious feeling were marked by racial boundaries. Alongside their successes in banking, finance, and other professions, upper-middle-class Jews who were proud of their "Germanness" were nonetheless assigned racial attributes that were presumed to be immutable. These usually included being cunning, imitative, materialistic, dishonest, egotistical, oversexed, disease prone, and given to conspiracy or revolutionary intrigue. Marr's book, *The Victory of Judaism over Germandom,* "changed the image of Jews from being a small, weak group to that of a world power."[3]

A scientific classification of human "types" based in a hereditary determinism that preceded the new genetics was well under way. Prior to this time, Jews had been bothersome and untrustworthy outcasts, religious underlings ever wandering Eurasia with no sense of place and marked with the stigma of having killed Christ. With the addition of a biologically based racial antisemitism, permanent and pathological bodily difference was taken as scientific "fact." As was the case for other "races," anthropological measurements of skulls supposedly reflected the mental and moral dimensions of race, and the above-average size of Jewish skulls was often taken to reveal evil cunning and/or a disposition to mental illness. Jews were thought to be inherently dangerous biological beings, infiltrating the nationalist pride of their presumed "superiors," and using their innate cunning to conspire to control the world. As racial antisemitism took hold, the very success of newly assimilated Jews made them even stronger targets for discrimination.[4]

As I describe in the last chapter, after 1890, pragmatic and political antisemitism was perfected in Lueger's Vienna. The immensely charismatic Lueger linked long-standing religious prejudice to new economic threats brought by industrialization and used political antisemitism strategically to manipulate the dissatisfied and economically threatened lower middle classes. By the end of World War I, despite the brave patriotism of many Jews in war, death and economic devastation were accompanied by the resurgence of a multidimensional antisemitism anchored in Catholic myth, enhanced by economic deprivation, and fully undergirded by the power of "rational" racial science.

All three of the men whose overlapping scientific story I have told were of Jewish ancestry. Tandler was raised in near poverty, his Jewish mother and father having moved from Iglau, Moravia, to Vienna when he was about three. Kammerer and Steinach, twenty years apart in age, were raised in assimilated Jewish families. Their fathers were established middle-class men. Steinach came from a medical family in Hohenems, a small Alpine town on the Austrian-Swiss border; his grandfather and father were both physicians. The town's history suggests that his family reflected the enlightened liberalism of assimilated bourgeois Jews.[5] Of the three, only Kammerer was born in Vienna. His father was German and his mother Jewish. He was attracted to the status and influence of the elite upper-middle-class intelligentsia, and outwardly he may have been surprised to be considered Jewish at all. All three men enjoyed professional respect, though Steinach and Kammerer worked in an institution of minor prestige, whose reputation was controversial (and Jewish). Each man had either converted from Judaism or identified himself as *konfessionslos* (an atheist). Tandler and Kammerer both married outside of the faith; and none of the three actively practiced any religion. But by 1920, religion was not the issue. They were perceived by their culture to be racially Jewish.

The effect of antisemitism would have been intensified by their varying, but shared attempts to promote a biology of flexibility that, by 1925, challenged the hereditarian rigidity that underlay fixed racial boundaries. What impact did their Jewish heritage have on their scientific success and their ultimate failure? It would be naïve to suggest that they failed because they were Jewish. And I do not. Genuine scientific questions were at stake, questions that many well-meaning individuals tried to answer on the basis of reasonable evidence. But scientists and their science are not immune to prejudice. In an atmosphere of great scientific uncertainty, the conceptual and empirical gaps in science combined with personality and the prejudices of the time to enhance their failure. As fascist ideology gained ground, by 1930, their view of biological flexibility had been stereotyped as a "racial physiology," erroneously propounded by

Jewish scientists and with no foundation in serious science.[6] They built it, critics argued, to reject their immanent inferiority. On this view, the rigid biology of race meant they had to be wrong; their own racial inferiority predisposed them to bad, antiracial science.

Historian Robert Proctor writes: "What has not been sufficiently appreciated is the extent to which the genetics of Mendel, Morgan, and Weismann was itself supported for political reasons."[7] In interwar Austria, antisemitism was an important part of that politics. To explore its impact, in what follows I integrate several strands of the scholarship on Viennese Jewish identity into a framework that assesses the impact of antisemitism on the fates of Kammerer, Tandler, and Steinach. They were an important part of the historical contribution of assimilated Jews to what scholars have called the Central European late-Enlightenment, a tradition that rested on combining social activism with the promise of reason and progress through natural science. With all of its devastation, the postwar period also brought hope for the return of what Stefan Zweig had described before the war as the Jew's "Golden Age of security," now extended to a wider swath of humanity in the Austromarxist program.[8] To varying degrees, each man shared this quest for unity and progress in a society torn by scientific debate and nationalist tension. And each eventually saw the exclusion of their contributions from the professional communities they had joined to realize that goal. The Jewish crisis of assimilation, which was first apparent in the 1880s, reemerged amid a postwar antisemitism strengthened by the authoritative dress of the new genetics. Kammerer, Tandler, and Steinach responded by challenging the rigidity of genetics to instead stress biological flexibility that, at least for Kammerer, explicitly rejected rigid racial boundaries.

An antisemitic caricature of Steinach (the most internationally famous of the three) appeared in a book presented to guests at a 1931 Viennese ball for journalists and authors. In it, Steinach is sketched alongside Sigmund Freud and Albert Einstein as one of the three most prominent Jewish scientists (fig. 8.1). An antisemitic poem that accompanies the sketch portrays the three as men who sought to remake the world and "push themselves in anywhere."[9] The assumption of their conspiratorial mentality was clear, and this was probably the majority perception of Tandler and Kammerer, as well. I emphasize Steinach less in my discussion of the impact of Jewishness on their failure because he played little direct role in the debates about heredity and because there are fewer personal sources with which to build a case for the impact of his Jewish heritage. Indeed in 1924, when Benjamin suggested to Steinach that some of the negative reaction to Steinach's work in America might be due to antisemitism, Steinach couldn't see it. He replied that the agitation against his work by the influential Chicago physician Morris Fishbein, who was Jewish, was not due to antisemitism.[10] Perhaps not; but as the caricature suggests, in the late 1920s,

Vienna was steeped in antisemitic reactions to Jewish scientists. Steinach, Kammerer, and Tandler could each be seen as racially inferior Jews building their science on flawed "Jewish" themes. In this context, antisemitism and their reactions to it played important roles in the fate of their synthesis of endocrinology and genetics; it surely contributed to their failure.

FIGURE 8.1. "Einstein—Freud—Steinach." Sketch from a book of caricatures that was given as a party favor to the ladies at a ball in Vienna, February 1931. Steinach was one of the three most prominent Jewish scientists.

Reproduction after a sketch in the Vorarlberger Landesarchiv, Bregnanz. Reprinted in Ernst Freud, Lucie Freud, and Ilse Grubrich-Simitis, eds., *Sigmund Freud: His Life in Pictures and Words* (Toronto: Penguin Books, 1985), 258–259.

"Germanness" and Antisemitism in Science and Medicine

Academic antisemitism was especially tinged with the "modern" flavor of race science, and by the 1880s it was rampant among faculty and students at the University of Vienna. In 1875, the renowned surgeon Theodor Billroth published a book in which he warned of the deterioration in the standards of medical education that had resulted from the large numbers of Jewish students then entering medical school in Central Europe. His racial justification found a large following among non-Jewish students at the university, who considered this view, by then based in science, to be modern and enlightened. Tandler and Steinach were medical students at the university, and the antisemitism of students and faculty most probably affected them. More than Kammerer, who was roughly ten to twenty years younger, they must have later felt the puzzlement so evident in Jewish memoirs of their generation that the culture, which they had eagerly embraced and which seemed so open to them, had turned on them.[11]

In Austria, as in Germany, a *Gymnasium* education was the gateway to entering the university and to membership in the elite professions of medicine, law, journalism, and finance. The Gymnasium experience was dominated by *Bildung*, the moral and cultural cultivation of character through education that was considered to be the foundation of the German national character. With it came a dedication to *Deutschtum*, a kind of mental and cultural Germanness, which at the beginning of the nineteenth century invoked in Austrians and Germans the German Enlightenment traditions of emancipation, equality, and tolerance.[12] It was an elitist educational system that, beginning at about age ten, excluded students in the lower and lower-middle classes. In contrast to those entering a Gymnasium, these children went to separate schools until the age of fourteen, when their educations ended. This educational segregation was not based only on financial considerations—many stipends were available—but on a structurally embedded aristocratic and academic elitism that perpetuated status barriers. Non-Jewish Gymnasium students were from homes of prestige, and many of their fathers had also attended Gymnasien. Their entry was horizontal. In contrast, "Western" Jews of Tandler, Kammerer, and Steinach's generations were exceptions. They used the system to change their social status, to sometimes enter vertically, and, in the process, they came to have a huge impact on Viennese modernism.

But the *Gymnasien* offered more than the possibility of sought-after upward mobility, acceptance, and prosperity. They also helped those of Jewish heritage establish an inner sense of their Germanness. Classical language and literature, humanistic reason, and the German language dominated the Gymnasium curriculum. The traditional Jewish regard for formal learning and a solid grounding in the German language enabled petty bourgeois and even a few lower-class Jewish sons to hope that the Gymnasium experience could be their means toward complete assimilation. For many, it sort of was. Way out of proportion to their representation in the population, they became "primary movers and shakers of modern Central European culture by the 1890s—whether as physicists, chemists, biologists, psychoanalysts, sociologists, journalists, composers, novelists, playwrights, and literary critics . . . secular assimilated liberal Jews increasingly set the cultural tone."[13] But success increased hostility and the perceived threat, and, as several scholars note, true assimilation was an unattainable ideal. Even in mixed schools, Jewish students remained marginal and socially isolated from their non-Jewish classmates. They were constantly reminded that they were Jewish. Nonetheless, they internalized the values of Bildung, which represented not just the means to see themselves as part of German culture but also a way to become Germanic agents of that culture.[14] Their education instilled a kind of stateless cognitive nationalism: a nationalism not of a polity, but of ideas and language that expressed "German" values of character, human obligation, reason, and morality. Especially in the multinational Habsburg state, Jews could be mental nationalists by virtue of their immersion in the values of Bildung.[15]

Jewish Cosmopolitanism and the Crisis of Assimilation

Several historians have stressed the unique historical situation of the fin-de-siècle Habsburg and post-Habsburg Jewish intelligentsia as distinct from that in Germany. Jewish intellectuals made outstanding contributions to both cultures in the nineteenth century, but important political and cultural differences framed distinct trends. The late nineteenth and early twentieth centuries were a time of surging European nationalism. Arguably, until the transition of Austria-Hungary from an empire to several successor states, the least nationalist European state was the Habsburg state. Compared to the strong nineteenth-century German emphasis on a Teutonic heritage of national "blood" purity, the multinational diversity of Austria-Hungary tolerated a range of identities. As racial nationalism grew elsewhere, the multinational flavor of the Austro-Hungarian Empire helped the Austrian Jewish intelligentsia become heirs to an Enlightenment tradition that championed equality and universalism. Enlightenment tolerance and universalism had surely declined in Austria with the failure of liberalism in the 1880s. But the fin-de-siècle Viennese Jewish intelligentsia kept these values alive and reframed them, even as they dealt with the crises of assimilation produced by the postwar rise of race science and nationalism.[16]

Though not technically liberals (many were socialists), many Viennese Jewish thinkers emphasized the values of human equality and universal human rights. With little sense of national boundaries, assimilated Jews, especially, accepted the transnational promise of an imagined post-imperial world. Historian Malachi Hacohen argues that the attempt to promote progressive transnational values championing diversity in search of human universals—cosmopolitanism—arose in the context of intense ethnonationalist debates and a profound and unbroken acceptance by many in Vienna of the German intellectual tradition. For assimilated Viennese Jews at the end of the nineteenth and the early twentieth centuries, "true *Deutschtum* was enlightened cosmopolitanism."[17]

Paradoxically, this Central European cosmopolitanism enhanced by Austrian multinationalism nonetheless privileged German language and culture. It contradicted the quest for universal humanity, because at least implicitly, it assumed the hegemony of Germanic education, language, and thought over those of other cultural traditions—particularly the Slavic peoples and the more religious Eastern Jews, who were immigrating to urban centers like Vienna in ever-larger numbers. As such, assimilated elite Western Jews endorsed a silent nationalism that challenged their intellectual and personal lives when, near the end of the century, pan-Germanist and racial antisemitism surged in the very culture on whose intellectual base they had founded their universalist impulses. In his analysis of the dilemmas of cosmopolitanism, Hacohen distinguishes a scholar's "cultural and political identities from his universal intellectual self." By 1920, this dilemma divided the identities of many highly educated Austrian Jews who advocated internationalism, but whose intellectual selves were linguistically, ideologically, and morally, if not essentially, German.[18]

The paradox of cosmopolitanism provides an excellent framework within which to explore the reform motives of Paul Kammerer and Julius Tandler and to ask how each was influenced by his perceived or personal sense of Jewishness. Both had been raised in families for whom a Gymnasium education had produced cultural advantage, and each identified with the linguistic and cultural tradition of Bildung as the spiritual and rational foundation of character. Tandler's family was "more Jewish" and more recently arrived in Vienna. And because of their education, their sense of scientific universals, and their tepid religiousity, neither man probably felt especially Jewish. But the stigma was there. They were continually reminded of it, as they were increasingly separated from the majority culture—marginalized while seeking an ideal of assimilation that was not possible.

Further, the scientific issue they addressed was in no way peripheral to debates about Jewish identity or the viability of assimilation. Many issues about the character and identity of the Jewish people after 1880 turned on how one saw the relationship between nature and nurture. How had Jews remained constant over the centuries as Jewish people had interfaced with so many different cultures? Was complete assimilation into the Germanic (or any other) ideal possible? Would it compromise treasured Jewish traditions? Or were the differences so fundamental that a separate Jewish state was the only solution?

Mitchell Hart quotes Arthur Ruppin, a Berlin statistician and a Jew, to show that Jewish thinkers were deeply involved in these debates; for them too, the relationship between nature and nurture was "the fundamental problem."[19] By 1920, in Vienna, the birthplace of Zionism, Tandler's position as a key player in Austromarxist reform and Kammerer's persuasive (to some) zoological demonstrations placed the two men in the middle of the issue at a time of resurgent antisemitism. Their notoriety and, especially for Tandler, their initial success very likely produced a backlash. The growth of race science, which by 1910 was anchored in major advances in German-speaking science, turned on them, and the culture that had shaped their characters now deemed them unfit to represent it. This was particularly paradoxical for scientists of Jewish ancestry, whose very expertise told them that they were outsiders. Hacohen refers to an "ethnonational marginality" fed by the universalism of the Jewish intelligentsia that made them Vienna's "non-Jewish Jews."[20] The correlate for many in science was the internalization of a sense of flawed Jewish science that made them Vienna's non-scientific scientists. Neither Kammerer nor Tandler could have failed to notice that he was being marginalized by a culture that by the mid-1920s was steeped in a dangerous mix of religious, political, and racial antisemitism. In different ways, they resisted the science that justified it and, in the process, they were altered by it.

Jewishness and Tandler's Response to Antisemitism

Julius Tandler's position as director of the Anatomical Institute and dean of the medical faculty placed him in a position of power at the University of Vienna. His Jewishness colored that power in his scientific and in his political career. He was considered by some to be "more Jew than anatomist"; in 1910, his appointment as Zuckerkandl's successor was feared by the antisemitic press as a sign of the "judification" of the University of Vienna.[21] Indeed, there were two anatomical institutes at the University of Vienna. Tandler headed one; and Ferdinand Hochstetter directed the other. The institutes represented different ideologies as well as different visions of the relationship between anatomy and other branches of biological science. Left-thinking and Jewish students, who went into medicine in large numbers, were inclined to join Tandler's institute; conservative, antisemitic and *Deutsch-völkish* students gravitated to Hochstetter's. Student-organized antisemitic attacks on Tandler's institute began somewhat regularly in 1923. They were probably fed by Billroth's critique, and their existence mirrored the academic antisemitism expressed toward many Jews then being educated at the university. That year, students forced their way into Tandler's institute, shouting that they were "seeking Jews." Tandler faced them, asking what they wanted. Not recognizing him as head of the institute, one student demanded Jews. When Tandler retorted that he *was* a Jew and head of the institute, they turned and left. This was one of several times in the decade when fascist students aggressively disrupted Tandler's institute. They shouted antisemitic taunts during lectures by his assistants; and they incited brawls. On a day in the late 1920s when angry students disrupted Tandler's anatomy lectures, Sigmund Freud's diary noted "antisemitic disturbances" at the university. Incensed by the need to physically defend his operation, Tandler had complained to the administration and requested measures to protect his students and assistants. The worst attacks came in March and May of 1933, as students encouraged by the Nazi rise to power in Germany started riots that injured several assistants and demolished areas of the institute.[22]

Tandler had converted to Catholicism in June 1899. Just a year later he married Olga Klauber, a Protestant, in a Protestant church. Occasional parish records, however, list him as *glaubenlos* (without belief). His biographer suggests that his motives for conversion were linked less to any religious beliefs and more to a desire for assimilation that could bring professional advancement.[23] In a university in which many positions had been closed to Jews, he held a position of unusual power, and he probably thought of himself less as a Jew and more as a man of science. But assimilation was impossible after 1900; he was considered a Jew. As his political career grew, Tandler faced similar prejudice from the critics of his health reforms. The reforms he instituted for the youth, for example, were considered by antisemitic detractors among the Christian Socials to build groups that bred "Jewish filth" among the young.[24]

Historian Klaus Hödl suggests that Tandler's double role, both concerned welfare reformer and Jewish scientist, only marginally shaped his personal identity. The medical community in Vienna had been heavily influenced by Jews who had gone into medicine in droves. Many of them accepted the scientific validity of the reliance on biological notions of race; and they were doubly subject to the prejudices it promoted.[25] Their responses varied from self-hatred to a radical Zionism that itself drew on the concept of a Jewish race. Tandler seems subject to neither. His diary describes the day, when, as a seven-year-old on Yom Kippur, he lost his faith. But he recontacted his Jewish identity rather late. In 1933, while traveling to China as a near exile, he wrote of his impressions as he passed the Red Sea and Mount Sinai. They brought back memories of his family, and of the hope that deliverance through the concept of Jerusalem had put in their hearts at the Seders of his youth. In tones of resignation, Tandler noted that the fate of his people was evident even then in Ahasver, the wandering Jew, who was visible over the waves of the sea: "he hovers like the ordeal of the Jewish people, inexhaustible and irreversible." Tandler too, was now wandering, perhaps anticipating his ejection from Austria and the people he had tried to help. He had neither sought nor desired the identity of a wandering Jew; but he sensed it now as part of his remembered past, part Jewish, part Viennese, and much a man of science. He may have seen how much the fate of the Jewish people was becoming his fate; he too, would be forced to wander.[26]

Hödl has compared Tandler's race hygiene to the eugenic approach that grew up among a tradition of authors committed to Jewish religious traditions and to their Jewish identities. He concludes that Tandler's approach to race hygiene took more the form of the non-Jewish camp, which after the war increasingly stressed negative eugenics. Hödl interprets Tandler as rejecting the preventative measures that Hödl sees as the hallmark of Jewish racial hygiene.[27] I disagree. Though, as Hödl stresses, Tandler placed much less emphasis on the eugenic value of Jewish religious tradition per se, his strong preference was for an approach to eugenic improvement that deemphasized selection and stressed the developmental potential of the young. Tandler heralded a scientifically conceived preventative medicine, one anchored in social reform, and he based his reforms on his version of productive eugenics, an approach that privileged children's development to enhance individual quality and forestall future hereditary damage. While the approach was not strictly preventative for some, it was so and more for others. He believed it would be restorative, as it offered a non-selectionist approach to positive correction and an enhanced human morality. In this sense, it was constructively preventative for society as a whole, a dimension he stressed over and above its value to the individual. By means of the changes to heredity brought via improved environments in one generation, the inheritance of acquired characteristics eliminated problems and yielded productive adaptations for the future.

There is also in Tandler the tension that emerges in the contrast between his origins in immigrant poverty and his rise to a position of great power and authority. His family was poor, poorer than those of Steinach or Kammerer in their youths. But at his peak, he rose to great influence, reaching something close to an intellectual ideal of power and influence. Did this instill true empathy for the plight of the poor? Sablik, his biographer, suggests yes, and he quotes several who agree.[28] Tandler's enormous postwar effort and the risks he took to achieve it reveal genuine concern for working families and their children. He hoped that others could rise as he had. But his authoritarian approach also implies that they must do it as he had, with his values, on the rational terms of enlightened reason and science understood through a Germanic ideal of character. This was his Jewish-cosmopolitan dilemma. The values of Deutschtum, often divorced from his Jewishness, would dictate what constituted a healthy moral character. His self-presentation smacks of cultivating and even enjoying the role of "Uncle Tandler," as the people sometimes called him.[29] And his ambiguous references to dangerous inferiority may reflect some of the shared prejudice that he acquired with his rise.

Though clearly identified with the progressive politics of the interwar period, Tandler was probably a bit less "German" and more "Jewish" than

Kammerer, who had probably been raised a Catholic. Tandler had in one generation moved further from poverty to power; as a youth he had rankled at his Gymnasium education, doing poorly, especially in the key language study areas.[30] As the above episodes illustrate, as an adult he was branded with the prevailing stereotypes of Jews. By converting, he distanced himself from some of the psychological and social burdens of "being" Jewish in an antisemitic culture. But his motive was to become more German, not less Jewish. And after a time in the postwar period, his success probably made the latter less visible to him. He aimed first and foremost at being a scientist who had significant social impact. For this reason, his internationalism was less a sociopolitical internationalism, and more the universal rationalism of science.[31] But, as his diary suggests, he may have retained more of his Jewish heritage than he liked to admit. He could hardly reject his history, especially given its forced acknowledgment by his culture. To do so would be to become someone else; and his musings at age sixty-four show that that had not happened; he too, was "a non-Jewish Jew."

Desperation, Antisemitism, and "Jewish" Neo-Lamarckism

Paul Kammerer's move to explore an endocrine explanation for somatic induction came at an increasingly difficult time in his life. Around 1910, in his late twenties, he was living the good life. He had married a baroness, Felicitas Wiedersperg, and for a time the couple lived amid music and various animals in the family home of his wealthy wife. In 1909, his research won the prestigious Soemmerring Prize from the Senckenberg Natural Science Society in Germany. But his passions, both personal and political, eventually came to haunt him. He had contracted gonorrhea, a problem that periodically recurred and that he later confessed to his American friend Benjamin. Just before the war, Erwin Baur, an important German geneticist, had challenged some of his zoological work. And changes were occurring in the scientific priorities of the Vivarium, where he worked. Despite the collaboration with Steinach, his work was increasingly peripheral to the laboratory's mission. His boss was concerned as early as 1921 that Kammerer had become a liability; the meetings of the lab's curatorial committee at the Austrian Academy of Sciences show that they were uncomfortable with his presence. The war and the economic crises that followed aggravated his problems. He had begun to talk of leaving Vienna as early as 1920; in the same year, he complained of "thoughts of escapism" (*Weltfluchtgedanken*), and of a rashness and impulsivity that, he said, had plagued him since puberty.[32]

In principle, Kammerer's values and his politics were quite consistent with a Jewish cosmopolitanism juxtaposed, as Hacohen suggests, with Bildung's privileged Germanness. Perhaps more so than for Tandler, transnationalism, tolerance, even a resistance to the intense nationalism of Germany rested comfortably in him alongside the hegemony of German language and culture. But the dilemma potentially produced by his mental Germanness was overshadowed by personal instability. A far greater dimension of his identity was his emotional insecurity, engendered by mental, physical, and financial troubles, which by the mid 1920s were probably fed by his increasing sense of being considered a Jewish outsider.

Kammerer had been at odds with his scientific critics and competitors well before the midwife toad scandal. He wrote with two voices, in a language appealing to the layman and worker as well as in a formal scientific language. The first was almost evangelical in tone, broadly speculative in an era of deepening specialization, and too exaggerated for the scientific standards of the day. It was passionately and openly political, a point often stressed by his detractors. Believing that bias was inevitable in science, Kammerer made his clear, attacked the racism and nationalist biases of others, and then arrogantly seemed to expect the superiority of his morality to prevail. But this placed him in an uncomfortable position, one mistrusted even by supporters of the inheritance of acquired characteristics. If the oneness of subjective and objective meant that all science carried social taint, could one then build an argument based in part on an appeal to scientific evidence that favored one view over the other? This was Kammerer's dilemma. His arguments for reform appealed to scientific logic favoring the validity of symbiosis and the inheritance of acquired characteristics, while at the same time implicitly undermining its objectivity. There are ways out of this dilemma; but Kammerer neither offered a solution, nor likely understood one. Instead, his political and ethical passions covered a contradiction that many found unsettling.

This tendency to overextend seemed to increase as he saw that his zoological work had become insufficient to win decisive support for the inheritance of acquired characteristics. His penultimate scientific hope was the proposed endocrine mechanism of somatic induction, the interstitial cells, and the heat rat evidence supporting it.[33] But this took Kammerer from the socially and scientifically contested area of race into equally problematical debates about sex and gender. And it did not persuade those critics whose problem was not so much the idea or its mechanism as the "race" or politics of those who championed it. As his life circumstances grew more extreme and his world crumbled, he perceived the shifting ground on which his key passions, productive eugenics and organic technology, rested. His emotions boiled, and he lived in extremes. In 1921, Kammerer resigned his Vivarium position, and he was formally released in November 1923. He was without a job in the midst of an extreme financial crisis.

Kammerer's correspondence with Benjamin in those years reveals a depressed and ill man, desperate for money. He began frantically looking for jobs in Europe and America. In 1922, Benjamin wrote that it should be easy to find teaching positions in the western United States, if nothing better turned up.

But by 1923, nothing had. And after separating from his wife and more troubles with Steinach, Kammerer told Benjamin that he wanted to leave; he would even "crawl to America as a servant (e.g. as your lab 'assistant') or a dishwasher," anything that his "weak heart" would let him do. He admitted to more suicide attempts in 1923: "I must enter the new world as a newer, freer man without the shadows that want to follow me from the old world. . . . Otherwise, I run the danger of losing my life, [a life] that is still of some meaning for the fulfillment of a few obligations. In order to appear not make empty talk, I confess to you that on the 3rd and 4th of July I twice tried suicide."[34]

He was a vain man, unabashedly promoting himself in ways considered embarrassing for a scientist. As the episode with Tandler and Kammerer's request for the photograph suggests (Chapter 7), some colleagues recognized and disliked this. Opposing his application for a professorship at the University of Vienna (see below), the influential paleontologist Othenio Abel described him as a *Reklammensch* (self-promoter), a view shared on the other side of the Atlantic.[35] T. H. Morgan referred to the "advertising" that often surrounded Kammerer's work, when he refused to be on the scientific committee that welcomed Kammerer on his arrival in New York.[36] During his six-month American lecture tour in 1923 and 1924, Kammerer regularly appeared in large headlines in the *New York Times:* "Biologist Predicts Critics' Conversion." The *Times* quoted him as planning new work that would "make American biologists' hair stand on end." When asked about American women, he responded, "Will never get enough of them. Do you get that?" Other pieces presented Kammerer as a victim. One headline read "Dr. Paul Kammerer, Associate of Steinach, Driven from Vienna by Poverty."[37] Kammerer was feeding the press and his style was brash, vain, and defensive.

Przibram, Kammerer's Austrian boss and sometime defender, confirms the insecurity behind that vanity. His memorial, written just after the suicide, gives us insight into the contradictions and burdens that defined Kammerer's character. Przibram noted that Kammerer's volatile temperament and his craving for recognition (*Geltungsdrang*) made it hard for him to get over small setbacks. By the mid 1920s, as he anticipated a move to Moscow, Przibram said, Kammerer was overcome "as oftentimes before by a heavy mental depression."[38] Przibram speculated that life's barbs had only deepened Kammerer's need for admiration. Not wishing to claim that Kammerer had solved the problem of the inheritance of acquired characteristics, Przibram nonetheless referred to slander. He noted that those who wished to alter the race exclusively through natural selection opposed Kammerer's vision, challenging his ideas not with science but with suspicions about the experimenter.[39] Kammerer became bitter and withdrew.

For an insecure and ill man, whose life's work was losing its scientific standing, the antisemitism present in his culture was likely to have eaten at his vanity, increased his bitterness, and fed his depression. Well before the scandal or his turn to endocrinology, antisemitism played a role among some of his critics, and may have fed his downfall. In 1918, a University of Vienna commission charged with considering Kammerer's application for the title of associate professor (*Extraordinarius*) discussed concerns about the quality and integrity of his work. The title would have conveyed great prestige, and the initial recommendation was positive. But an unusual delay separated the recommendation and the commission's final decision. Several members changed their minds when Kammerer's book, *Das Gesetz der Serie* (The law of series), was published after the positive recommendation had been submitted. The commission was asked to reconvene, and, though some members considered the popular book irrelevant to the assessment of his science, others took it as an affront to the integrity of all science. To the critics it spoke to Kammerer's unscientific attitudes, and this group prevailed in the final decision. The application was rejected.

Several irregularities surrounded the process, and they have led both historians and some of Kammerer's contemporaries to charge that the denial was based on personal dislike, political differences, and antisemitism. Rejecting the assertion that the decision was made purely on scientific grounds, Peter Berz and Klaus Taschwer interpret the politics behind the official report (which referred largely to the book). At least one commission member—the paleontologist Abel—later wrote that fighting "the threatening judification and foreign takeover of the University of Vienna" had been a first priority after the war. Historian Albrecht Hirschmüller too, mentions that antisemitism surrounded the rejection of Kammerer's application, but he offers little evidence. He instead stresses criticisms of Kammerer's popular works, which critics believed "biologized" social issues. But biologism was everywhere in conservative accounts of race and degeneration at the time, so it just as likely that the objectionable nature of Kammerer's attempts stemmed from his politics and his half-Jewish heritage. Newspaper accounts of the episode denying the promotion state that Kammerer had even predicted that to be approved he would have to certify his "purely Aryan roots," a task that his mother's Jewish ancestry made impossible.[40] That he saw antisemitism as a part of the threat of racism is clear in Kammerer's attack on Houston Stewart Chamberlain, the British-German race theorist, German nationalist, and Darwin hater. Despite his many logical inconsistencies, Chamberlain's wide appeal, Kammerer wrote, was due to his talent for interpreting life in antisemitic terms.[41]

Antisemitic slurs were sometimes added later by figures who at first had cautiously endorsed Kammerer's zoological work, but then condemned it. German evolutionist Ludwig Plate was one whose initial concerns were, by 1933, framed in racial tones: "All of these reports by Kammerer the Jew must be taken with extreme skepticism."[42] Never in question were the political and social views of Kammerer critic, Berlin anatomist Hermann Stieve. In the 1920s, Stieve had been extremely critical of Tandler and Steinach's work on the secretory function of the interstitial cells and its link to somatic induction (see Chapter 4). After his move to Berlin in 1935, Stieve's research no longer focused on animals; he examined the anatomical changes associated with the cessation of ovulation in the reproductive organs of recently executed women who had lived for months under the fear of death. The women were Nazi prisoners, and one can only imagine the extreme stress of their final days. Given his conservative views on gender and his later career, it is likely that antisemitism played a large part in Stieve's rejection of Kammerer and the "Vienna school" in the early 1920s.

Antisemitism also played a role in Kammerer's American reception. Some of the most negative views of Kammerer in America came from the Johns Hopkins biostatistician Raymond Pearl. In 1923, Pearl wrote neurologist Henry Donaldson that he had evidence that Kammerer was a fraud, but the evidence had been lost in a 1919 fire in his laboratory. Contacts between the two men appear to have been minimal, but Pearl's antisemitism was a part of a concerted American attempt to keep Jews out of American universities in the 1930s. Pearl, who was in correspondence with several other Kammerer critics, including Morgan, Bateson, and zoologist Howard Parshley, presented the issue as a moral one—Jews were, he assumed, less moral than Gentiles—and as a biological struggle for existence, which Gentiles must win. Especially given Kammerer's attempts to seek employment in the United States, it is likely that antisemitism played a role in Pearl's reaction, too.[43]

But the clearest and most flagrant expression of antisemitism came from the influential German geneticist Fritz Lenz. Lenz was the first Central European professor of racial hygiene at the University of Munich in Germany. A student of August Weismann, he became convinced early on that there was no such thing as the inheritance of acquired characteristics. Munich was the center of an approach to race hygiene anchored in a rigid biology of race. And Lenz increasingly used inflexible heredity as a justification for Nordic racial superiority. He moved up quickly in German academia. From 1913 to 1933, he was editor of the *Archiv für Rassen- und Gesellschafts-Biologie,* and in 1921, he coauthored the first edition of the most authoritative work on the new genetics in the German language, a book that became the Nazi's reference work on the genetics of a fixed racial hierarchy.[44] In 1933, he was called to a professorship in Berlin.

On one issue, Kammerer's cosmopolitan egalitarianism flew directly in the face of Lenz's politics. Lenz linked his rigid racial hierarchy to mental as well as bodily differences, which included a hereditary disposition to crime in "negroes" and the definition of Jews as a "mental race." Their bodily features did not mark Jews as clearly as did their "racial" minds. As a race, he said, Jews were oversexed, mimetic, exploitative, and witty. Moreover, Lenz believed that

Jews had a biological tendency to accept Lamarckian interpretations of heredity. In his view, the idea itself was a racial attribute.[45] In several places, Lenz specifically mentioned Kammerer as an illustration of this innate Jewish tendency toward neo-Lamarckian thinking. Proctor quotes the 1927 edition of the genetics textbook, in which Lenz explained that he understood why Jews might want to become Teutonic Germans: "If acquired characters could be inherited, then, by living in a Teutonic environment and by adopting a Teutonic culture, the Jews could become transformed into genuine Teutons. . . . But . . . the Lamarckian doctrine is an illusion. . . . Jews do not transform themselves into Germans by writing books on Goethe."[46] The oversimplification expressed at the end of the statement was a common ploy to ridicule the scientific complexity of ideas about somatic induction.

Because Lenz's views on gender were even more extreme than his views on race, it could not have helped that Kammerer's revised scientific effort based developmental flexibility on malleable views of masculinity and femininity through Steinach's reproductive endocrinology. In 1929, Lenz reiterated the Jewish racial source of the neo-Lamarckian "illusion" in a review of the film *Salamandra*, which had been made by the Soviets to melodramatically present Kammerer as a scientific martyr. Lenz's review is more character assassination than review. He reprinted his criticisms, originally published in 1913, and he not only recounted the midwife toad revelations (suggesting that a replication would have been better than suicide if Kammerer had really wanted to redeem his honor) but he also implied conspiratorial political maneuvering. He even reviewed Kammerer's failures with women, which he attributed to his (Jewish) psychopathology: hysteria. Oversexed, hysterical, mimetic, dishonest, conspiratorial, and exploitative: it reads as if Lenz were presenting Kammerer as a case study in Jewish racial weakness. Indeed, his title refers to "the case" of Kammerer, which, he said, was "psychologically so interesting." Three years after Kammerer's death, there was no longer any need to defame the man, despite the exaggerated flavor of the film. With very few exceptions, whatever persuasive power he had once had was gone. But Lenz was not arguing, he was illustrating. The Kammerer "case" supported the fact of a Jewish "mental race" with its flawed tendency to accept flexible heredity (which, he wrote, was also apparent in the thinking of Przibram, Iltis, and Goldscheid). In 1924, Hugo Iltis had written Lenz to dispute the view that an acceptance of neo-Lamarckism in heredity was a Jewish racial attribute.[47] Lenz rejected Iltis's argument, and his racial stereotype of Kammerer first surfaced when, as he refuted Iltis, he added that Kammerer really was a Jew, despite his father's German side. As he later described the Kammerer "case," Lenz wrote: "For a half Jew the wishes which dispose one in the direction of the 'inheritance of acquired characteristics' are if anything much stronger."[48] Hybrid or not, the stereotype bred true; it even increased.

As Lenz shows, for those so inclined, Kammerer made an easy target. Remembering him in a 1951 interview conducted by his son, Kammerer's friend Iltis noted Kammerer's vanity and said that some colleagues ridiculed him "because of his 'dandy' type of behavior." But Iltis also said that Kammerer was persecuted because of his Jewish ancestry and his political beliefs. The "sentiments" that were to be Hitler's were prominent among many at the University of Vienna in the 1920s, said Iltis. "To them K[ammerer]. his views and his activities were a ready target like the red flag is to a bull." Those objections, already strong in the late teens, were strengthened by Kammerer's (and Iltis's) explicit rejection of the racism that motivated them. Kammerer was being undermined by the very concept that his passion challenged: race.[49]

Divergent Reactions

Tandler and Kammerer reacted very differently to the antisemitic attacks on their reputations and their science. The differences were shaped by their different histories, their different habits of mind, and their distinct personalities. Tandler's childhood history of poverty and his rise via confident self-determination were crowned by great faith in the rationality of science. His background in clinical medicine enabled Tandler to resolve the uncertainty that dominated the liberal scientific culture of interwar Vienna with the empirical reality of the diseased individual. Compared to the abstract and general character of statistical and theoretical statements, the patient was the lived reality of medicine—concepts and laws were useful, but just abstractions (see Chapter 6); they mattered less than action that could transform the health and well-being of individuals and of society.

Kammerer's philosophical idealism, his monistic worldview, was anchored in a oneness that he could see everywhere. Nature and mind, genes and environment, nation and universe, science and art: all were one. The contrast between Tandler's pragmatic approach to reform and Kammerer's idealistic and philosophical one is apparent in how the two men dealt with race. Kammerer challenged the concept on ethical and biological grounds, eventually siding with Hertz and Boas, as his monistic worldview fused biology and society so that they fed one another (see Chapter 5). This made race biologically flexible and socially transient. By contrast, when Tandler referred to race, he treated it as a foggy statistical abstraction, not meriting much discussion. Referring to the difficulties with the concept of race, Tandler wrote that sometimes "race" is nonchalantly forced on us to name a group of individuals. But "conceptual confusion and the misuse of words are the order of the day" and the labels were "mostly nothing other than an expression of currently missing knowledge on the cardinal attributes of race."[50] He neither redeemed nor rejected the race concept; rather, because science had not yet spoken, he avoided it and argued instead that race hygiene was condition hygiene (see Chapter 6). His strong claim was not about race; it was about how to proceed scientifically to enact change. The conflict between universalism and Germanic intellectual hegemony could be resolved, in Tandler's mind, by the transcendent objectivity of science. Social reality was practical, not philosophical.

This and their different personalities produced quite different reactions to the antisemitic attacks. Tandler responded only when necessary to protect his students and assistants, and he did so by appealing to authorities. If the attacks wounded him personally, which is doubtful, he did not react publicly. By contrast, Kammerer's sensitive temperament led him to overreact: to become defensive and respond with exaggerated and overdrawn conclusions. Though he rarely referred to antisemitism in the rejection of his work, both moral revulsion and personal bitterness probably motivated his attacks on racism. By contrast, when antisemitic attacks were not disrupting his laboratory and endangering his students, Tandler's Jewishness could be almost invisible to him, at least until near the end. He certainly saw the racism that Kammerer noted after 1918. But Tandler cultivated first and foremost the identity of a scientist; he wielded and respected the authority of objective science, which, personal considerations aside, would prevail over cultural tradition and bias, Jewish, German, or otherwise.

Antisemitism and Failure

Antisemitism by no means produced the failures of these scientists. Priorities, ego, insecurity, vanity, rapidly changing and irreconcilable types of scientific evidence, fractures produced by emerging new disciplines, vague new constructs like "gene" and "hormone," all played a role. It is important also to note that the varying influences affected the lives and careers of these three men differently at different points in time. Their convergence in 1910, when the Vivarium was new, Kammerer was rising, and Tandler and Steinach had just begun their research on the interstitial cells was nothing like their convergence in 1928, after Kammerer had died and when Tandler wrote "Gefahren." And at no time did one influence prevail over the others. But it would be difficult to deny that each man was subject to antisemitic prejudices that affected the reception of his science. They shared a belief in developmental flexibility that challenged the rigidity of genes at a seminal moment in the history of biology; they championed the power of a new and for some difficult construct, hormones, to ensure the scientific foundation of change; and each was of Jewish ancestry at a time when race was for very many a valid concept in heredity. By 1929, others may have agreed with Lenz that the inheritance of acquired characteristics was a flawed Jewish idea. It was probably not so treated, at least not in the same way, in 1880, when the debates between August Weismann and Herbert Spencer drew attention to a challenging new scientific problem. As the scientific and political trends shifted in the 1920s and early 1930s, both racists and nonracists, racialists and nonracialists, were influenced by the changes. Several scientists of Jewish heritage also rejected the inheritance of acquired characteristics, and some Jewish thinkers accepted a rigid concept of race.[51]

But the political changes in Central Europe and the expulsion and eventual murder of Jews make it clear that racist genetics undergirded by Weismann's assumption of inflexibility could (but did not always) justify racial antisemitism. Noting the eventual twentieth-century triumph of Weismann's view, Proctor writes: "Political motivations can be as important in justifying correct views in science as they are in justifying false views." He uses the chilling example of Theobold Lang's article "National Socialism as a Political Expression of Our Biological Knowledge" (written in 1930). Nazism, wrote Lang, was based on solid biological principles; one of them was the "immutability of the human genetic material."[52] These biological principles remained solid scientific conclusions until after the turn of the twenty-first century. For reasons that did not originate in antisemitism, early twentieth-century science had begun to change, posing important and challenging problems for those supporting flexible heredity. Among them were the decline in teleological frameworks, the rediscovery of Mendel's laws, the formation of new disciplinary boundaries, and debates about mutation and the particulate nature of heredity; none of these should be construed as latent antisemitism. Prejudice did not replace science; but neither was it founded on pseudoscience. Rather, at the time the two became, for many reasons, interwoven as part of the justification for Central Europe's collective social demand for a rigid concept of heredity; and "rational" science became a foundation for virulent racism.

9

Asymmetry, Failure, and Flexible Heredity

In his retrospective summary of the fifty-year history of the science of genetics written in 1950, Richard Goldschmidt celebrated the triumph of Mendelism. In the process. he used two phrases to dismiss the likes of Paul Kammerer, Julius Tandler, and others, who—heretically by 1950—had seriously explored the inheritance of acquired characteristics, attempting to reconcile it with chromosomal cell division and Mendel's laws. Lamarckism, Goldschmidt said, reflected an "historical oddity" of the young science; it was one of "the childhood diseases of genetics" in its youth.[1] By contrast, he compared the impact of "mature" genetics to that of Galileo; he praised not just neo-Darwinism, but "hyper-Darwinism"; and he could see progress in it—even in 1950—as the foundation of selectionist eugenics. Those years of debate in the teens and twenties seemed misspent in 1950, as Mendelian ratios and population genetics fused with cellular genetics and neo-Darwinism to unify biological science in what is often called "the Modern Synthesis."[2] It would only get better in 1953, when James Watson and Francis Crick presented the double helix as a model of the chemical structure of genes.

It would be hard to overstate how many outstanding advances in genetics and evolutionary theory after 1930 warranted Goldschmidt's pride. It *was* a triumph, and Kammerer's, Tandler's, and Steinach's attempts were overcome in part by the success of that triumph. In his groundbreaking study of alternatives to Darwinism at the turn of the century, Peter Bowler writes, "In the end, the [Lamarckian] theory was abandoned not because it lacked proof, but because Mendelian genetics proved so much *easier* to elaborate into the conceptual foundation for the study of heredity."[3] But the triumph of genetics was not solely a scientific one—success was based in more than concepts, data, and disproof. And the scientific advances made were not without cost; its deserved success should not blind us to the ways in which genetics directly and indirectly contributed to the abuses of eugenics.

The rejection of the inheritance of acquired characteristics was an important aspect of how genetics fed selectionist eugenics. In Central Europe, the scientific advances that helped spawn its success were intertwined in complex matrices of new scientific discovery and pressing social questions, which, in that historical moment, seemed to threaten humanity to its core. In the episode I have presented, scientific debates took place in an historical context that altered the outcome, as the mix of science, society, and prejudice with debates about progress, degeneration, and the problem of teleology fueled a momentum of success that overwhelmed the endocrine-based hypothesis of somatic induction. Historical accounts of the transition must include assessments of the social forces that helped sustain one view over another and the ways in which those forces were challenged or facilitated by societal values and reactions. The failures of Kammerer, Steinach, and Tandler were partly personal. Kammerer's suicide, Tandler's flight amid revolution and the rise of fascism, and Steinach's exile, isolation, and near ridicule helped bury their respective scientific advances. Yet personal failure need not lead to scientific failure; more was involved as their science, too, failed.

The Asymmetry of Success and Failure

It is a commonplace that history is written by the winners. Because the fruits of science are so obvious to us, this privileging of the present may be even more tempting in the history of science. Recent advances in our understanding of the relationship between, science, scientists, and society have produced important counterweights to the temptations of success.[4] Historians of science now avoid assuming that the scientific concepts dominant in their own time would, but for a weak or incomplete scientific base, eventually emerge from the past. It is trivial to say that we know today that Steinach was right about the interstitial cells; by the standards of today's science, they are the major source of the mammalian sexual secretions, testosterone and estradiol. And the possibility that hormones might (somehow) alter genes is a good idea by today's standards. Hormones like estradiol activate receptors, which act as "transcription factors" that control gene action. But Kammerer, Tandler, and Steinach did not and could not have known these things; and they in no way anticipated them. In their time, there were no hormone receptors, there were no transcription factors, there was not really DNA, at least not as a recognizable concept in heredity. Their understanding cannot be imposed on today's concepts. But success still has great pull. How then do we understand the complexities of the shifting momentum that culminated in their failure?

I have proposed an asymmetry between success and failure that disproportionately favors success as scientific momentum shifts (see Chapter 1). Exploring this asymmetry can help clarify what, by the standards of an earlier time, facilitated the overly early failure of a disputed alternative. A moment's thought might suggest that such asymmetry would weight failure more than success. Ideally, science disproves. In principle, disproof takes very little, and there are lots of long-abandoned ideas and conclusions. Aristotelian and Ptolemaic astronomy are gone; the solar system is heliocentric, not geocentric; natural theology could not explain extinction; rather, species evolve, and they often disappear. But a more fine-grained examination that encompasses the day-to-day unfolding of scientific practice suggests the opposite: asymmetry favors the momentum of success. The macro failures that occur at the level of what Thomas Kuhn called paradigm shifts can obscure the short-term momentum of success. That momentum accelerates at a micro level, within working research traditions, sometimes not driven by theory or worldview (which may surely play a role). Less obvious and perhaps more powerful in the early stages of a transition are the dimensions of workaday science that shape and are often shaped by societal trends. These smaller changes enable a little good evidence to produce more good evidence like itself.

The asymmetry that favors the successful turn can fuel the synergy between science and society: more and more of the same good and useful "findings" become more relevant to the social applications and reforms that favor them. Success affords more reform, even broader success. Especially as scientific materials and methods become commercialized, those adopted in the wake of initial success reinforce the replicability and power of the relevant "findings." In the practice of science, accelerating success can be fed by new materials, experimental setups, "model" organisms, community networks, new journals, and arcane terms that beget more terms understood and used only by a smaller and smaller circle of experts.[5] Leading institutes get imitated as their findings prevail and their leaders come to dominate. The predictive value of their methods, the internal consistency of their empirical work—a consistency that feeds upon itself as terms and method become unquestioned foundations—breed more of themselves and look ever more "scientific."

I have presented interwar Vienna as a time during which the intertwined matrices of the sciences of heredity with sociopolitical issues were especially dynamic. The infusion of scientific debate with sociopolitical problems made the changing momentum altered by the "internal" sources of success even greater. That convergence may not always be so extreme. But in interwar Vienna, the greater significance of Tandler's and Kammerer's failure is that, at the time, debates about the validity of the inheritance of acquired characteristics were imbedded in a mix of race science, racism, sexism, and the tensions between democratic socialism and emerging fascism. The concept was certainly honestly questioned; but by 1915, there did not yet exist a predominance of success or failure on either side of the debate. In 1940, and certainly by 1950, when Goldschmidt wrote, the picture looked very different. By the 1960s, when the genetic code relating DNA base triplets to single amino acids was deciphered, the success of genetics soared. But the memory of success should not distort the pull of the ideas that failed. In Vienna in the 1910s, the playing field of competing assumptions and the values associated with the evidence favoring one side or the other was more nearly equal. It is, I suggest, in just such periods, when momentum has not yet built, that the weight of evidence on either side is most vulnerable to social influence that, intertwined with the trappings of materials and scientific infrastructure, can disproportionately favor success.

As momentum shifts and success becomes ever more successful, failure can be rapid. Journals, new terminology, disciplinary boundaries, commercialization of materials converge with societal trends to accelerate the process, as findings are endorsed by larger and more powerful social and political concerns. The result for the losers is not necessarily a gradual and cumulative logical failure. Concepts and data disappear, are shouted down, rejected, or quietly abandoned—but not because they were, in some meaningful use of the term, wrong.[6] Arguably, two things make the case I have presented a bit more noisy and revealing of the process: the political attention given at the time to the implementation of Red Vienna; and the intensity and melodrama—some of it self-generated—surrounding Kammerer's fall. Their visibility helps us see how the eventually asymmetrical outcome was affected by interlocking scientific and societal influences, as scientific debates about the relationship between nature and nurture took place amid great social change.

Patterns of Failure

This case of failure is not representative. Few if any are. But the failures of Kammerer, Tandler, and Steinach can reveal patterns that might also come into play in other periods of scientific transition. These patterns may or may not involve what we take to be the transcendent features of modern scientific logic and method. This is not to say that such features are unaffected by social influence; rather, at some times they may be less affected than at others by the uniqueness of time or place. For example, the elaborate system of producing and maintaining *Drosophila* mutants in Morgan's genetics laboratories rapidly perpetuated findings favoring one side over the other.[7] The power of technique, language, replicability, and standardization was immense. By contrast, few could quickly build on the subtle, nonstandardized techniques that produced Kammerer's carefully nurtured toads and salamanders and Steinach's chronic rat implants. Similar 'structural' differences might affect scientific transitions occurring in other historical contexts. By contrast, other dimensions had more resonance with the social issues that characterize this historical period. Between 1890 and 1930, for example, debates about progress and the implications of a blind versus a purposeful view of evolution were especially significant in Central Europe. For this reason, despite their search for mechanism, the issue of teleology contributed to Tandler and Kammerer's failure, as "organic technology" was lumped with teleological forms of neo-Lamarckism that were unacceptable to experimental science. Despite no real difference in philosophy, what in that context was perceived as the more "modern," but not more mechanistic, view of heredity gained greater momentum.

These patterns represent interlocking interfaces of discourse through which opponents form competing matrices that unite science and society on each side of the dispute. Interwar neo-Lamarckians like Kammerer and Tandler were no less political than those whose race science they challenged. Their science too, was part of a sociopolitical matrix; it was not a set of purely rational or empirical concepts challenging a tainted fusion of science and protofascism. For this reason, the failure of their ideas was influenced as much by the decline in the egalitarian and socialist views with which they were aligned as by the value of their evidence. Especially in times of transition and debate, such patterns of interconnection may alter the shifting fortunes of the contenders. In what follows I summarize a few that are apparent in the case I have presented.

First, the interlocking interfaces may occur at many levels. One is the individual scientist. Kammerer made himself an easy target for critics who—sometimes for scientific reasons, sometimes not—did not like his ideas. Tandler, though a target, was by no means an easy one; if anything, he equivocated to appeal to many sides.

A second level is that of the community or thought collective of which a scientist is a part. Despite the similarity of their ideas, it would be a mistake to place Kammerer and Tandler in the same scientific collective. Anatomy and constitutional medicine had some conceptual, but little community overlap with amphibian zoology; the two men were positioned in very different professional circles, and they might have been more effective had this not been the case. The threads linking them were their similar use of reproductive endocrinology and their visions of the utility of the inheritance of acquired characteristics. Much else separated them (see Chapter 7).

A third level is represented in the sociopolitical affiliations that link scientists with political positions. Both Tandler and Kammerer were committed socialists who were associated, respectively, with the Austromarxist and Austrian monist networks. They shared ideas, but their views engaged somewhat different social supports, an important exception being Rudolf Goldscheid (discussed in Chapters 5 and 7). Consideration of these various levels can reveal considerable diversity among the advocates of a given concept that can lessen their common impact.

Multiple levels are complicated by ongoing change. At each level the interlocking interfaces within and between competing positions move, shift surely, and react. Tandler and Kammerer's turn to reproductive endocrinology in the midst of debates about nature and nurture and the issue of somatic induction represents such a move. Earlier approaches had not addressed the problem of mechanism, and effective reaction to critics required a shift that surely added new directions and new problems, notably the interstitial cell question (see Chapters 2 and 4). It is also likely that the initial success of Red Vienna, understood by both sides of the debate in part as a rejection of the new genetics, served as an impetus to the postwar rise of Reche's right-wing race hygiene (Chapter 7). Some success on what becomes the losing side can increase counterreaction on the other side. There were surely other influences that strengthened Reche's response (those internal to anthropology, for example); but it is important to recognize that competing matrices move in a dance of reactivity, often manifest (or presented) as a purely scientific debate.

The debate surrounding the interstitial cells illustrates another influence that made it difficult for Tandler and Kammerer's hypothesis of somatic induction to gain momentum. This relates to discipline formation. The validity of use inheritance was fundamentally an issue in the study of heredity. Tandler, Kammerer, and Steinach attempted to solve it with a concept from the new field of endocrinology. Only one man of the three, Steinach, can be considered an expert in either field. As disciplines shifted and the fields of genetics and endocrinology formed, Kammerer and Tandler were both working outside of their main areas of expertise. When disciplines take sides on debates that unite each within, they can lose a shared foundation in the methods, language, and concepts needed to resolve the issues that fall between them as specialization provides less and less common ground. A key issue on which the endocrine hypothesis of somatic induction depended was whether or not the interstitial cells could secrete hormones (Chapter 4). But this question had no resonance within genetics. As physiologists challenged cytologists on the origin and function of the cells, proponents of the new genetics had no basis on which, if they heard it at all, to judge the issue. Disciplinary boundaries had become more distinct, and this made it easier for one or both groups to tacitly delegitimize the voices of outsiders, voices that could be ignored by the increasingly specialized audiences to which they were directed.

In his analysis of the failure of Aramis, a 1980s Parisian experiment in automated personal rail transit, Bruno Latour distinguishes between the proximate and the remote causes of failure. He acknowledges that each contributes to the outcome; but remote causes carry more weight. The proximate is "a last straw," while the remote cause is built "into the nature of things."[8] His distinction is relevant here. Antisemitism was a structurally remote and pervasive dimension of Austrian culture; its origins surely preceded the rise of the modern sciences of heredity. And by 1900, though it had already entered science, antisemitism was reinvigorated as a structural underlay of Lueger's Vienna as much for economic and political as for scientific reasons (see Chapter 7). The same might be said in Austria of the pan-German values of Bildung and the debates about progress and purpose. They influenced the debates among the scientists, but their origins were remote to the scientific problems of the time. Kammerer's suicide, which had an impact, was, on the other hand, a proximate influence, one of many which by the mid-1920s could have precipitated decline, once the more remote sources of success had begun to gather momentum. Kammerer's death tipped; but it did not create. Vanity and suicide could influence those already affected by antisemitism, disciplinary divisions, or the hold of mechanism. But they were irrelevant for others, like Tandler, for whom, "mangled" toads notwithstanding, other dimensions carried the day (see Chapter 7). Kammerer, Tandler, and Steinach were individual pressure points on the side of a debate advocating developmental flexibility and flexible heredity. Steinach brought scientific prestige and a cutting edge new approach in endocrinology. Kammerer brought broad conceptual integration, a passion for the vision offered by the key concept, and a body of visible and novel experimental work. Tandler brought power and the reality of Red Vienna as an example of the concept's promise, arguably made more significant by his scientific prestige. Their efforts engaged powerful remote influences working against them, and neither vanity nor linguistic hedging had much impact, once the other side gained momentum.

Guilt by Lamarckian Association: Mechanism and Hereditary Predeterminism

One remote and pervasive influence deserves further analysis as an example of how a persistent issue can be used to new ends in a particular historical context. In interwar Vienna, the problem of environmentally induced and adaptive novelty was embedded in debates about the role of teleology in biological explanation. Kammerer and Tandler's ideas were presented as the question of teleology was once again being debated in Central European science, and Kammerer chose his phrase "organic technology" specifically to align his support for use inheritance with a mechanistic view of causation. Despite this stance, however, others who defended mechanism sometimes rejected his views by lumping them with those of scientists who used the concept to argue for teleology, avoiding the important differences. Examining the role of teleology in debates about flexibility in heredity illustrates how philosophical and conceptual issues can be co-opted by the other side to misrepresent opponents and enhance their decline. During heated debate, scientists can exploit sensitive issues to accelerate failure through guilt by association.

Carl Detto's distinction between somatic and parallel induction had been an attempt to resolve the problem of teleology and show the consistency between causal-mechanistic accounts and the possibility of somatic induction (see Chapter 3). But it also opened the door to interpret—as Weismann already had—any demonstrated effects as simple cases of parallel induction. For many ardent selectionists who acknowledged only parallel induction, linking utility to structure without teleology required an acceptance of germ-plasm–based predeterminants—rigid genes—that inherently ensured an adaptive result. As long as an unseen "hereditary disposition" in the corresponding germ-cell determinants was triggered by an external change, parallel induction prevailed; there would be no somatic induction and no problem of teleology. Accepting only parallel induction eliminated both, but at the cost of a rigid hereditarianism.

However, as proponents more and more drew on physiology to explain use inheritance, denying somatic induction became harder. Some opponents of flexibility in heredity resorted to a categorical rejection of virtually all "Lamarckism" on philosophical grounds. Heinrich Ziegler, a Weismann student and proponent of a nationalist and liberal biology of society, illustrates the link between predeterminism in heredity and the philosophy of mechanism. He also illustrates how readily all neo-Lamarckians (including Kammerer) could be summarily rejected and counted as outmoded teleologists by those who recognized only parallel induction. In 1910, Ziegler wrote a summary of the debate between Weismannism and Lamarckism—these were his terms. Ziegler went to great pains to move biology away from vitalism and show that mind, purpose, and God had nothing to do with evolution. He relied on what he acknowledged were unseen changes in the germ-plasm to argue that selection just sorted among outcomes to perpetuate preexisting potentials already present in the germ. In defending mechanism, he also decried the teleology of those less disposed to predeterminism in particles of heredity. After quoting the original Lamarck, who was hardly the basis of the debate by 1900, and citing August Pauly, Ziegler referred to the "cell mysticism (*Zellenmystik*)" to which contemporary neo-Lamarckism had degenerated.[9]

Selectively citing Detto, Ziegler interpreted all demonstrations of the inheritance of acquired characteristics as instances of parallel induction. He argued that any changes produced by environmental alteration were sustained, not by inherited acquisitions of direct adaptations in parents but by generations of hidden selection manifest in their offspring. And along with Pauly, whose account was teleological, Ziegler also challenged Kammerer, whose explanation shared none of the "mystical" properties of those proposing psychic and volitional processes in evolution.[10] Ziegler maintained that the midwife toad results occurred because the larvae had retained an innate predisposition to slower development. The presumption of predetermined latent pathways preexisting in the germ eliminated adaptive novelty based in flexible heredity and the need for a mechanism of somatic induction. When novelty appeared, a hidden genetic predisposition (*Veranlagung*) had simply been reactivated.

But Kammerer's turn to physiology challenged Ziegler's philosophical framing of the debate. Kammerer presented his "organic technology" as an approach by which humans could intentionally mold the future of the species through a biology that was devoid of immanent purpose. With an explanation in physiology, he, like Tandler, intended a Lamarckism based in a mechanistic account of how interacting antecedent causes—the chemistry and physiology of hormones and the energies of the external environment—produced their effects. The hormones of the double gonad were philosophically consistent with Ziegler and Detto: physiology had no immanent purpose. Hormones were the material channel (*Leitungsbahn*) to which Detto had referred, and with such an account no residue of teleology need cling to the inheritance of acquired characteristics. In *Neuvererbung*, Kammerer even addressed the philosophical problems and confusions introduced by versions of neo-Lamarckism that drew on mentalism and intentionality to explain the inheritance of acquired characteristics. He wrote, "The danger [of vitalism] lies near, and the school of the 'psycho-Lamarckians' (for example, A. Pauly) has given over to it." To avoid any teleological implications, he suggested replacing mental metaphors such as "memory" with the concept of organic persistence (*Beharrungsvermögen*). "Through this our biological thought becomes closer to physicalistic thinking." Kammerer could not yet claim to offer the entire physical picture; but he believed that heredity must be explained physically, and he wrote, "nor are we far removed from also being able to explain it just physiologically."[11]

So, far from endorsing teleology, as Ziegler charged, Kammerer was trying to do the opposite. But he also sought to challenge Ziegler's predeterminism and replace it with developmental flexibility (see Chapter 4). Truly novel adaptations could originate in individual adjustment in ways mediated by cellular chemistry and endocrine physiology. Novelty was bought by human wisdom harnessing biological flexibility, the foundation of productive eugenics, not by teleology. Ziegler's rejection of adaptive novelty in any form illustrates how intertwined philosophical debates about mechanism versus teleology were with the social and scientific debates about heredity and their relation to eugenics.[12] For those for whom novelty brought unwanted or dangerous social change, assuming a system of hereditary predeterminism made it easy to avoid. To do so, however, Ziegler and others had to reject Kammerer's mechanistic account as just another attempt to bring teleology back into biology. American plant ecologist

Henry Cowles illustrates the effect of such misrepresentation. In a review of Detto's book, Cowles portrayed "the Lamarckian" as one who "conceives of a vital mechanism that looks out for the future, and holds advantageous reactions in readiness for conditions which have never yet occurred." He too rejected all Lamarckism as teleological, noting that teleology was a dead issue in America: "In this country, at least, there is no need for such a continuous and hearty lampooning of teleological and vitalist views [as Detto refuted], for they have been long since abandoned by most scientific investigators." Any neo-Lamarckians that complicated the issue could be cast with the taint of outmoded teleology and quickly dismissed.[13]

Kammerer's and Tandler's solution represented a third alternative that relied neither on a teleological and purposeful view of biology nor on a rigid hereditary predeterminism. The endocrine system provided a physicalist account of adaptation that at once avoided "volitional moments" and saved flexibility. Seen as the mediators of somatic induction, the products of the interstitial cells challenged both the reliance on predeterminism and the exclusivity of parallel induction. And both men were convinced that the approach could replace the rigid notion of heredity that defined much postwar eugenics, positive or negative. Their physiological accounts of somatic induction offered a means by which society could be improved through an ethic of social reform reflected in the promise of Red Vienna. Neither Tandler nor Kammerer developed a successful mechanistic neo-Lamarckism. But their efforts represented a plausible attempt anchored in pioneering changes in reproductive endocrinology, which some critics lumped with teleology in an effort to silence them.

Their failed attempt to integrate hormones and heredity to account for somatic induction offers a revealing historical episode in the interface of science and society. It helps us resist the temptation to read history backward and assume that the idea had to be rejected because it was wrong. That was not the case at the time. The alternatives presented by Tandler, Kammerer, and Steinach were not outdated versions of Lamarck bathed in the shadow of teleology and uninformed by contemporary changes in early twentieth-century science. They were not pseudoscientific, fraudulent, or unfounded in evidence. Rather, they represented an avant-garde approach to an old question newly understood by the concepts of reproductive endocrinology. Their attempts, though incomplete, were scientifically plausible and debated in high-level scientific circles. They show us the conditions under which competing research traditions can be affected by social and political change, and they add to a body of historical work on the interwar period, which shows that the science of the time did not inevitably lead to the drastic outcomes that followed Hitler's rise after 1933. Rather, there was another side that raised serious ethical concerns and scientifically sound alternatives. But as the situation worsened, a socioscientific and political matrix that was moving toward entrenched biological racism silenced their voices. The mix can be construed neither as the fault of science nor as a scientific conspiracy. It was instead, a reciprocal pattern of facilitation whose success was ensured by the silencing of dissenting voices in science. Robert Proctor again clarifies: "The moral, then, is not just that the Nazis supported science or that science supported the Nazis [and their predecessors], but that there was a struggle that led to the triumph of racial Nazi policy, and this struggle was played out, at least in part, in the spheres of science and medicine."[14] Tandler's, Kammerer's, and Steinach's attempts to integrate endocrinology into a view of flexible heredity were part of that struggle, and their stories help us navigate the complex ways in which the plausible and evidence-based in science can be overcome. They help us understand how scientific failure encompasses and transcends science.

Epilogue

For almost fifty years modern genetics was dominated by the so-called "central dogma," framed by Sir Francis Crick in the 1960s. This is the idea that the flow of causal influence in heredity always moves from the genes outward to protein, organism, and environment, but not in the reverse direction, from environment, organism, or protein back to the genes.[1] The central dogma represents the mid-twentieth century rejection of the possibility that environmental influences could act indirectly, through the body, on genes. But, since the sequencing of the human genome in the early 2000s, scientists have increasingly questioned the central dogma. The discovery that there were not enough coding human genes to do what we had been told for decades that they were probably doing led scientists to explore new approaches, and the field of epigenetics emerged as a force arguing the opposite.

Epigenetics examines the developmental impact of cellular, structural, and environmental influences that alter gene action, but do not change the protein coding base sequences of messenger RNA. Epigeneticists have shown that features of the social and material environments related to poor nutrition, stress, and even child abuse can alter heredity from very early in development to produce attributes that are inherited across generations. And they are uncovering new and intriguing cellular mechanisms that explain how this happens and, therefore, how the presence of environmental influences may alter heredity to improve developmental outcomes. The development and evolution of adaptive structure, behavior, and cognition may even require such processes.[2] These challenges to the central dogma have made some scientists once again receptive to a kind of Lamarckian inheritance, now understood in terms that are consistent with the known chemistry of heredity.[3]

This reversal in our view of the relationship between genes and environment has brought us back to Kammerer and Tandler's basic question, a question that was nearly heretical just twenty years ago: what internal processes and environmental influences exist that adaptively (or maladaptively) alter heredity? Some have even suggested that Paul Kammerer was a forerunner of the changes now occurring, an issue about which there is debate.[4] The resolution of that debate is less important here than the fact that, after being consistently and definitively rejected by the majority of mainstream science, a failed concept can return, albeit completely transformed, to reassert the very point that once condemned it: the environment's formative role in affecting heredity to generate beneficial outcomes inherited by descendents.

Both success and failure are relative to an historical moment, and the return of a widely rejected but plausible scientific concept highlights the historical complexity of the failure of Kammerer's, Tandler's, and Steinach's vision in the 1920s. Their ideas and those of several others foreshadowed (but did not anticipate) the changes taking place in the study of heredity today. They were on a reasonable track. But they foreshadowed almost none of the important scientific details. These would be foreign to them. The terms and the spirit of their proposals, their scientific and their social base, were of a different time. But the recent changes help us see that in some sense, they did not fail as completely as it seemed they had from the point of view of the 1940s through the 1980s. That environmental change might be inherited was not a ridiculous idea, inconsistent with knowledge about heredity and condemned to unavoidable teleology. Especially when distinct scientific disciplines draw on one another's concepts to address a question as complex as the relationship between nature and nurture, socially digested accounts of the technical and empirical accomplishments of opponents rarely capture the complexities of the effort. The result can contribute to a cultural memory of scientific success that embellishes the advances and undervalues serious challenges that were once issues of intense debate.

The scientific debates about somatic induction in Central Europe in the 1920s are history. But their resonance with the recent changes in genetics makes it worth hearing their story. The idea of "flexible heredity" in any scientific dress raises difficult cultural and social questions. For example: if, as several investigators now suggest, epigenetic mechanisms greatly contribute to obesity and cancer, why is some research so concerned with applying and popularizing personal genome sequencing, so that each of us may know exactly what our individual coding DNA says? Why not pay more attention to the environments that organize and activate that DNA? How many group differences in health and well-being that are still linked to race can be traceable to environmental effects that alter gene expression early in development? These are modern versions of the kinds of questions that Tandler and Kammerer considered, as they applied their scientific arguments to social conditions.

Understanding the matrix of converging influences that occurred under troubled social conditions at one point in the history of a changing scientific consensus can help us see the complexities raised for another. Particularly as it engaged the social concerns embedded in the discourse about nature "versus" nurture, the failure of the endocrine hypothesis of somatic induction shows us that for those who prevail, success can depend on social and cultural influences as well as on scientific evidence.

NOTES

ABBREVIATIONS

APS	American Philosophical Society Library, Philadelphia, Pennsylvania
Duke	Duke University, Special Collections Library, Durham, North Carolina
IGM	Institut für Geschichte der Medizin (Department and Collections of the History of Medicine), Medical University of Vienna, Vienna
KI	The Kinsey Institute for Research in Sex, Gender, and Reproduction, Bloomington, Indiana
VGA	Verein für Geschichte der Arbeiterbewegung (Labour History Society), Vienna
NYAM	New York Academy of Medicine Library, New York
ÖAW	Österreichische Akademie der Wissenschaften (Austrian Academy of Sciences), Vienna
ÖNB	Österreichische Nationalbibliothek (Austrian National Library), Vienna
WBR	Wienbibliothek im Rathaus (Vienna City Library), Vienna
RA	Rockefeller Archive Center, Sleepy Hollow, New York
UWL	University of Wisconsin Library, Special Collections, Madison, Wisconsin

1. THREE FAILED SCIENTISTS

1. August Weismann, "The All Sufficiency of Natural Selection. A Reply to Herbert Spencer," *Contemporary Review* 64 (1893): 309–338, 321, quoted in F. B. Churchill, "The Weismann-Spencer Controversy over the Inheritance of Acquired Characteristics," in *Human Implications of Scientific Advance*, ed. E. Forbes (Edinburgh: Edinburgh University Press, 1978), 451–468, 458. See also Peter Bowler, *The Eclipse of Darwinism* (Baltimore: Johns Hopkins Press, 1983); F. B. Churchill, "August Weismann and a Break from Tradition," *Journal of the History of Biology* 1 (1968): 91–112; F. B. Churchill, "From Heredity Theory to Vererbung: The Transmission Problem, 1850–1915," *Isis* 78 (1987): 337–364; Robert Richards, *Darwin and Evolutionary Theories of Mind and Behavior* (Chicago: University of Chicago Press, 1987). On Weismann, see R. G. Winther, "August Weismann on Germ-plasm Variation," *Journal of the History of Biology* 34 (2001): 517–555.
2. Ian Hacking, *Rewriting the Soul: Multiple Personality and the Sciences of Memory* (Princeton: Princeton University Press, 1995), 16.

3. The view is associated with the ideas of the French biologist Jean Baptiste Lamarck and often described as "Lamarckian." But by 1880, "Lamarckian" and "neo-Lamarckian" approaches shared little with Lamarck's original theory. Both terms referred to the idea of the inheritance of acquired characteristics. For details of the scandal, see Sander Gliboff, "The Case of Paul Kammerer: Evolution and Experimentation in the Early 20th Century," *Journal of the History of Biology* 39 (2006): 525–563; and Arthur Koestler, *The Case of the Midwife Toad* (New York: Random House, 1971).
4. Jennings to Ross Harrison, January 21, 1924, H. S. Jennings Papers, Paul Kammerer folder, APS.
5. H. S. Jennings Papers; unpublished lecture notes, "Inheritance of Acquired Characteristics," no. 29a, 1924, 8, 14, APS.
6. H. S. Jennings Papers, unpublished lecture notes, no. 13b, 1920, 12, APS.
7. Herbert Spencer Jennings, *The Biological Basis of Human Nature* (New York: W. W. Norton, 1930), 345.
8. Jennings referred to figures. If they were from Kammerer's published accounts, he was using photos of some of the same specimens that were later declared frauds.
9. Chandak Sengoopta, "Glandular Politics: Experimental Biology, Clinical Medicine, and Homosexual Emancipation in Fin-de-Siècle Central Europe," *Isis* 89 (1998): 445–473; Chandak Sengoopta, *The Most Secret Quintessence of Life: Sex, Glands, and Hormones, 1850–1950* (Chicago: University of Chicago Press, 2006).
10. Karl Sablik, *Julius Tandler: Mediziner und Sozialreformer* (Vienna: Schendl, 1983).
11. "Government Arrests Viennese Professor," *New York Times*, March 18, 1934. Felix Czeike's entry for Tandler in the *Historisches Lexicon Wien* (vol. 3: 415) states that Tandler's efforts were an example that awakened the interest of the world.
12. Carl E. Schorske, *Fin-de-Siècle Vienna: Politics and Culture* (New York: Vintage Books, 1981); Steven Beller, ed., *Rethinking Vienna 1900: Austrian History Culture and Society* (New York: Berghahn Books, 2001).
13. Naomi Oreskes, *The Rejection of Continental Drift: Theory and Method in American Earth Science* (Oxford: Oxford University Press, 1999); J. Sant, "Wegener, Galileo and Darwin," scientus.org. Accessed September 2011: http://www.scientus.org/index.html.
14. Celia Roberts, *Messengers of Sex: Hormones, Biomedicine, and Feminism* (Cambridge: Cambridge University Press, 2007).
15. Robert Proctor, *Racial Hygiene: Medicine under the Nazis* (Cambridge, Mass.: Harvard University Press, 1988).
16. Written with Siegfried Grosz. See also Julius Tandler, "Konstitution und Rassenhygiene," *Zeitschrift für angewandte Anatomie und Konstitutionslehre* 1 (1913–14): 11–26.

2. REHABILITATING SEXUALITY

1. Harry Oosterhuis, *Stepchildren of Nature: Krafft-Ebing, Psychiatry, and the Making of Sexual Identity* (Chicago: University of Chicago Press, 2000), 250.
2. Georges Canguilhem, *The Normal and the Pathological* (New York: Zone Books, 1991); Ian Hacking, *Rewriting the Soul: Multiple Personality and the Sciences of Memory* (Princeton: Princeton University Press, 1995); Ian Hacking, *The Social Construction of What?* (Cambridge, Mass.: Harvard University Press, 2000). Eugenics was part of the trend; see Ian Dowbiggin, *Keeping America Sane: Psychiatry and Eugenics in the United States and Canada, 1880–1940* (Ithaca: Cornell University Press, 1997).
3. Daniel Pick, *Faces of Degeneration: A European Disorder, c. 1848–c. 1918* (Cambridge: Cambridge University Press, 1989); Gunter Mann, "Dekadenz—Degeneration—Untergangsangst

im Lichte der Biologie des 19. Jahrhunderts," *Medizinhistorisches Journal* 20 (1985): 6–35; Irmgard Mueller, "Degeneration als Prinzip des Wirbeltier-Phylogenese. Ein Kapital aus der Geschichte der Evolutionstheorie im 19. Jahrhundert," in *Miscellen zur Geschichte der Biologie: Ilse Jahn und Hans Querner zum 70. Geburtstag,* ed. A. Geus, W. Gutmann, and M. Weingarten (Frankfurt am Main: Waldemar Kramer, 1994), 117–135; Richard Soloway, *Demography and Degeneration: Eugenics and the Declining Birthrate in Twentieth Century Britain* (Chapel Hill: University of North Carolina Press, 1995); Annemarie Wettley, "Zur Problemgeschichte der 'Dégénérence,'" *Sudhoffs Archiv* 43 (1959): 193–212.

4. J. Edward Chamberlin and Sander Gilman, eds., *Degeneration: The Dark Side of Progress* (New York: Columbia University Press, 1985), xiii.
5. See Heins-Peter Schmidenbach, "'Ist nicht wirklich dieses ganze zersetzende Naturwissenschaft ein Irrweg?' Virchow und die Zellularpathologie," *Medizinhistorisches Journal* 27 (1992): 26–42.
6. Paul Weindling, *Health, Race and German Politics between National Unification and Nazism 1870–1945* [1989] (Cambridge: Cambridge University Press, 1993), 1.
7. Heide Rimke and Alan Hunt, "From Sinners to Degenerates: The Medicalization of Morality in the 19th Century," *History of the Human Sciences* 15 (2002): 59–88; Wettley, "Zur Problemgeschichte der 'Dégénérence'"; Ian Dowbiggin, "Degeneration and Hereditarianism in French Mental Medicine, 1840–90: Psychiatric Theory as Ideological Adaptation," in *The Anatomy of Madness: Essays in the History of Psychiatry, vol. 1,* ed. W. F. Bynum, R. Porter, and M. Shepherd (London: Tavistock, 1985), 188–232; Francis G. Gosling, *Before Freud: Neurasthenia and the American Medical Community, 1870–1910* (Champagne-Urbana: University of Illinois Press, 1987); Rafael Huertas, "Madness and Degeneration, Part I. From 'Fallen Angel' to Mentally Ill," *History of Psychiatry* 3 (1992): 391–411; A. Liegeois, "Hidden Philosophy and Theology in Morel's Theory of Degeneration and Nosology," *History of Psychiatry* 2 (1991): 419–427; Mueller, "Degeneration als Prinzip."
8. Pick, *Faces of Degeneration.*
9. Emil Kraepelin, "Zur Entartungsfrage," *Zentralblatt für Nervenheilkunde und Psychiatrie* 31 (1908): 745–751, 747. Soloway (*Demography and Degeneration*) notes that by 1900 the term "degeneration" referred to an irreversible hereditary state, while "deterioration" implied that environmental remedies could help. On Kraepelin, see Victor Roelcke, "Biologizing Social Facts: An Early 20th Century Debate on Kraepelin's Concepts of Culture, Neurasthenia and Degeneration," *Culture, Medicine and Psychiatry* 12 (1997): 383–403.
10. Kraepelin, "Zur Entartungsfrage," 747. Max Nordau noted that no matter how slight the degeneracy in an individual, its transmission would accumulate in descendents; see Rimke and Hunt, "From Sinners to Degenerates."
11. Roelcke, "Biologizing Social Facts." See also Eric Engstrom, *Clinical Psychiatry in Imperial Germany: A History of Psychiatric Practice* (Ithaca: Cornell University Press, 2003).
12. Sander Gilman. "Sexology, Psychoanalysis, and Degeneration," in *Degeneration: The Dark Side of Progress,* ed. J. E. Chamberlin and S. Gilman (New York: Columbia University Press, 1985), 72–96, 72. See also G. Robb, "The Way of All Flesh: Degeneration, Eugenics, and the Gospel of Free Love," *Journal of the History of Sexuality* 6 (1996): 589–603; Thomas Laqueur, *Solitary Sex: A Cultural History of Masturbation* (Cambridge, Mass.: Zone Books/MIT Press, 2004); and Roelcke, "Biologizing Social Facts."
13. By 1910, psychiatrists began to separate themselves from the notion of degeneration (Roelcke, "Biologizing Social Facts"). But World War I again strengthened the threat and the connections among sexuality, degeneration, evolution, and eugenics.

14. Oosterhuis, *Stepchildren*, 108. See also George Mosse, *The Image of Man: The Creation of Modern Masculinity* (Oxford: Oxford University Press, 1996); Jennifer Terry, *An American Obsession: Science, Medicine, and Homosexuality in Modern Society* (Chicago: University of Chicago Press, 1999); and Jeffrey Weeks, *Sex, Politics and Society: The Regulation of Sexuality since 1800* [1981] (London: Longman, 1989).
15. Julius Bauer, *Die konstitutionelle Disposition zu inneren Krankeiten* (Berlin: Springer, 1927).
16. Merriley Borell, "Organotherapy and the Emergence of Reproductive Endocrinology," *Journal of the History of Biology* 18 (1985), 1–30; Anne Fausto-Sterling, *Sexing the Body, Gender Politics and the Construction of Sexuality* (New York: Basic Books, 2000); Nelly Oudshoorn, "Labortests und die gemeinsame Klassification von Sexualität und Geschlecht," in *Die Experimentalisierung des Lebens*, ed. H.-J. Rheinberger and M. Hagner (Berlin: Akademie, 1993); Nelly Oudshoorn, *Beyond the Natural Body: An Archaeology of Sex Hormones* (London: Routledge, 1994); and Celia Roberts, *Messengers of Sex: Hormones, Biomedicine, and Feminism* (Cambridge: Cambridge University Press, 2007).
17. Cornelius Medvei, *The History of Clinical Endocrinology: A Comprehensive Account of Endocrinology from Ancient Times to the Present* (Pearl River, N.Y.: Parthenon, 1993); Chandak Sengoopta, *The Most Secret Quintessence of Life: Sex, Glands, and Hormones, 1850–1950* (Chicago: University of Chicago Press, 2006).
18. Sengoopta, *Secret Quintessence*, stresses the importance of clinical work in supporting the endocrine view.
19. See Medvei, *Clinical Endocrinology*; Thomas Schlich, *Die Erfindung der Organtransplantation* (Frankfurt: Campus, 1998); Sengoopta, *Secret Quintessence*; Chandak Sengoopta, "The Modern Ovary: Constructions, Meanings, Uses," *History of Science* 38 (2000): 425–488; see also J. G. Gruhn and R. R. Kazer, *Hormonal Regulation of the Menstrual Cycle: The Evolution of Concepts* (New York: Plenum, 1989).
20. See Adele Clarke, *Disciplining Reproduction: Modernity, American Life Sciences, and the Problems of Sex* (Berkeley: University of California Press, 1998). Clarke notes that in the United States reproductive endocrinology lagged behind for cultural reasons.
21. Deborah Coen, "Living Precisely in Fin-de-Siècle Vienna," *Journal of the History of Biology* 39 (2006): 493–523; Veronika Hofer, "Rudolf Goldscheid, Paul Kammerer und die Biologen des Prater-Vivariums in der liberalen Volksbildung der Wiener Moderne," in *Wissenschaft, Politik und Öffentlichkeit*, ed. M. Ash and C. Stifter (Vienna: Facultas Verlags und Buchhandels AG WUV–Universitäts Verlag, 2002), 149–184; and Wolfgang L. Reiter, "Zerstört und Vergessen: Die biologische Versuchsanstalt und ihre Wissenschaftler/innen," *Österreichische Zeitschrift für Geschichte* 10 (1999): 585–614. On solidism, see Sengoopta, "Modern Ovary."
22. Erna Leskey, *Die Wiener medizinische Schule im 19. Jahrhundert* (Graz: Hermann Böhlaus, 1965). On the Exner family, see Deborah Coen, *Vienna in the Age of Uncertainty: Science, Liberalism, and Private Life* (Chicago: University of Chicago Press, 2007).
23. Exner was an expert on physiological optics and the neural mechanisms of vision.
24. Sigmund Exner, "Physiologie der männlichen Geschlechtsfunktionen," in *Handbuch der Urologie*, ed. A. Frisch and O. Zuckerkandl (Vienna: Alfred Hölder, 1904), 208–265, 216. Despite his emphasis on sexual flexibility, Exner still distinguished the active sex drive of men from the passive reaction of women, ibid.
25. Ibid., 224.
26. On Artur Biedl (1869–1933), see Medvei, *Clinical Endocrinology*.
27. Physiologist A. A. Berthold castrated cocks in the 1840s, attempting to isolate the gonads. In one castrated bird, he removed the testes and repositioned one gland

near the intestines. The bird developed normally. When Berthold removed the transplanted grafts, the bird no longer showed interest in females, lacked aggression, and its comb and wattles faded and shrank. But other investigators could not replicate the work. Exner's own experiments with castrations and transplants failed, but his student, Artur Foges, continued them; see Sengoopta, *Secret Quintessence.*

28. Ibid., chapter 2.
29. Quoted in Artur Biedl, *Innere Sekretion* (Berlin: Urban und Schwarzenberg, 1910), 326; see also Susan Mosedale, "Science Corrupted: Victorian Biologists Consider 'The Woman Question,'" *Journal of the History of Biology* 11 (1978): 1–55; Sengoopta, "Modern Ovary"; Hans Simmer, "Zum Frauenbild Rudolf Virchows in den späten 1840er Jahren," *Medizinhistorisches Journal* 27 (1992): 292–319.
30. Cesare Lombroso, Guglielmo Ferrero, and Hans Kurella, *Das Weib als Verbrechin und Prostituirte* (Hamburg: Verlagsanstalt und Drückerei, 1894), 140.
31. A. K. Christensen, "A History of Studies on Testicular Leydig Cells: The First Century," in *The Leydig Cell*, ed. A. Payne, M. Hardy, and L. D. Russell (Vienna, Ill.: Cach River Press, 1996), 2–29. Rudolf Kölliker confirmed the presence of the cells in human males in 1853. In males, the terms "Leydig cell" and "interstitial cell" were synonymous, and Steinach regularly applied the term "interstitial cell" to females as well.
32. Tandler's early work on freemartins, genetically female cows that are masculinized prenatally by hormones from their male twin, preceded the more famous experiments done by the American Frank Lillie.
33. Julius Tandler, "Über den Einfluß der innersekretorischen Anteil der Geschlechtsdrüsen auf die äußere Erscheinung des Menschen," *Wiener klinische Wochenschrift* 23 (1910): 459–467; Julius Tandler and Siegfried Grosz, *Die biologischen Grundlagen der sekundären Geschlechtscharaktere* (Berlin: Springer, 1913).
34. Julius Tandler and Siegfried Grosz. "Über den Saisondimorphismus des Maulwurfhodens," *Archiv für Entwicklungsmechanik* 33 (1911), 297–302, 302.
35. Eugen Steinach, "Geschlechtstrieb und echt sekundäre Geschlechtsmerkmale als Folge der innersekretorischen Funktion der Keimdrüsen," *Zentralblatt für Physiologie* 24 (1910): 551–566, 564. Sengoopta, *Secret Quintessence*; Per Södersten, "A Historical and Personal Perspective on the Aromatization Revolution: Steinach Confirmed," in *Brain Aromatase, Estrogens and Behavior*, ed. Jacques Balthazart and Gregory Ball (Oxford: Oxford University Press, 2012); and Heiko Stoff, *Ewige Jugend. Konzept der Verjüngung vom spaten 19. Jahrhundert bis ins Dritten Reich* (Köln: Böhlau Verlag, 2003), provide excellent summaries of Steinach's work. See also Fausto-Sterling, *Sexing the Body*; Oudshoorn, "Klassifikation von Sexualität"; and Chandak Sengoopta, "Glandular Politics: Experimental Biology, Clinical Medicine, and Homosexual Emancipation in Fin-de-Siècle Central Europe," *Isis* 89 (1998): 445–473.
36. Eugen Steinach, "Willkürliche Umwandlung von Säugetier-Männchen in Tiere mit ausgeprägt weiblichen Geschlechtscharakteren und weiblicher Psyche," *Pflügers Archiv für Physiologie* 144 (1912): 71–106. Normal males, castrated males, and normal sisters served as controls that were raised with and were littermates of the transplanted animals.
37. Eugen Steinach, "Feminierung von Männchen und Maskulierung von Weibchen," *Zentralblatt für Physiologie* 27 (1913): 717–724.
38. This was especially apparent in the development of the penis. Compared to males that were castrated, those with successful implants had penises that "no longer deserved the name" (Steinach, "Willkürliche Umwandlung," 87).
39. Ibid., 104.

40. Steinach, "Feminierung von Männchen," 723.
41. Steinach, "Geschlechtstrieb und Geschechtsmerkmale," 566.
42. Steinach, "Willkürliche Umwandlung," 104. Males with a huge overproduction of interstitial cells sometimes even forced matings with unreceptive females, something that Steinach said did not occur in normal rats. On Steinach's discussion of homosexuality, see Florian Mildenberger, . . . *in der Richtung der Homosexualität verdorben. Psychiater, Kriminalpsychologen und Gerichtsmediziner über männliche Homosexualität 1850–1970* (Hamburg: Männerschwarm, 2002); Sengoopta, *Secret Quintessence*; and Eugen Steinach, "Histologische Beschaffenheit der Keimdrüse bei homosexuelle Männern," *Archiv für Entwicklungsmechanik* 46 (1920): 29–37.
43. Eugen Steinach, *Sex and Life* (New York: Viking Press, 1940), 5–6.
44. For Steinach's theory of hormonal sexual antagonism, see Sengoopta, "Glandular Politics," and Fausto-Sterling, *Sexing the Body* (chapter 6).
45. Eugen Steinach, "Pubertätsdrüsen und Zwitterbildung," *Archiv für Entwicklungsmechanik* 42 (1916): 307–332; Eugen Steinach, "Künstliche und natürliche Zwitterdrüsen und ihre analogen Wirkungen," *Archiv für Entwicklungsmechanik* 46 (1920): 12–28.
46. Steinach, "Pubertätsdrüsen und Zwitterbildung," 320.
47. Because sperm and eggs were also called "secretions," many scientists used the terms "inner secretion" or "incretion" to indicate the hormones.
48. Eugen Steinach and Guido Holzknecht, "Erhöhte Wirkung der inneren Sekretion bei Hypertrophie der Pubertätsdrüsen," *Archiv für Entwicklungsmechanik* 42 (1916): 490–507, 506.
49. Jane Maienschein, "What Determines Sex? A Study of Converging Approaches, 1880–1916," *Isis* 75 (1984): 456–480.
50. Richard Goldschmidt, "Experimental Intersexuality and the Sex-Problem," *American Naturalist* 50 (1916): 705–718. On Goldschmidt's enzyme hypothesis, see Garland Allen, "Opposition to the Mendelian-Chromosome Theory: The Physiological and Developmental Genetics of Richard Goldschmidt," *Journal of the History of Biology* 7 (1974): 49–92; Nathan Ha, "The Riddle of Sex: Biological Theories of Sexual Difference in the Early Twentieth Century," *Journal of the History of Biology* 44 (2011): 505–546; and Marsha Richmond, "The Cell as the Basis for Heredity, Development, and Evolution," in *From Embryology to Evo-Devo: A History of Developmental Evolution*, ed. Manfred D. Laubichler and Jane Maienschein (Cambridge, Mass.: MIT Press, 2007), 169–211. Steinach and Goldschmidt used different terms for the points on their continua. Steinach described his intermediaries not as "intersexual" but as "bisexual," a difference that may suggest his disregard of chromosomes.
51. Biedl, *Innere Sekretion*, 333.
52. Eugen Steinach and Heinrich Kun, "Antagonistische Wirkungen der Keimdrüsen-Hormonen," *Biologia Generalis* 2 (1926): 815–834; Eugen Steinach and Heinrich Kun, "Luteingewebe und männliche Geschlechtscharaktere," *Pflügers Archiv* 227 (1931): 266–278.
53. Steinach and Kun, "Luteingewebe und männliche Geschlechtscharaktere," 278, emphasis in original.
54. Knud Sand, "Moderne experimentelle Sexualforschung, besonders die letzten Arbeiten Steinachs ('Verjungung')," *Zeitschrift für Sexualwissenschaft* 7 (1920): 177–197, 179. Biedl, Sand, and Alexander Lipschütz all supported the hypothesis.
55. Steinach, *Sex and Life*, 87–88.
56. On Viennese crises of sexual identity, see G. Bischof, A. Pelinka, and D. Herzog, *Sexuality in Austria* (New Brunswick, N.J.: Transaction, 2006); David Luft, *Eros and Inwardness*

in *Vienna: Weininger, Musil, Doderer* (Chicago: University of Chicago Press, 2003); Chandak Sengoopta, *Otto Weininger: Sex, Science and Self in Imperial Vienna* (Chicago: University of Chicago Press, 2000).

57. Magnus Hirschfeld to Julius Tandler, August 9, 1929, Julius Tandler Papers, Correspondence, Department and Collections, IGM.
58. Herbert Steiner, ed., *Sexualnot und Sexualreform. Verhandlungen der Weltliga für Sexualreform* (Proceedings of the Fourth Congress) (Vienna: Elbemühl, 1931); J. Novak, "Innere Sekretion und Sexualität," in *Sexualnot und Sexualreform. Verhandlungen der Weltliga für Sexualreform*, ed. Herbert Steiner (Vienna: Elbemühl, 1931), 223–228, 226.
59. Steiner, *Sexualnot und Sexualreform*, xix.

3. PAUL KAMMERER AND FLEXIBLE HEREDITY

1. Undated draft of a review of Kammerer's *The Inheritance of Acquired Characteristics* (New York: Boni and Liveright, 1924) for *Medical Herald and Physiotherapist*, ms 45, Harry Benjamin Collection, Correspondence, Paul Kammerer Folder 3, KI.
2. Howard Parshley in *New York Herald Tribune*, November 16, 1924, ms, "Biology and Human Society," written for *American Review*, Harry Benjamin Collection, Correspondence, Paul Kammerer Folder 3, KI.
3. Hans Przibram, "Paul Kammerer als Biologe, *Monistisches Monatshefte* 11 (1926): 401–405, 402.
4. For details on Kammerer's life and the professional disputes, see Peter Berz and Klaus Taschwer, "Afterword," in Arthur Koestler, *Der Krötenküsser: Der Fall des Biologen Paul Kammerer* (Vienna: Czernin, 2010); Sander Gliboff, "The Pebble and the Planet: Paul Kammerer, Ernst Haeckel, and the Meaning of Darwinism," Ph.D. dissertation, Johns Hopkins University, 2001; Albrecht Hirschmüller, "Paul Kammerer und die Vererbung erworbener Eigenschaften," *Medizinhistorisches Journal* 26 (1991): 26–77; and Arthur Koestler, *The Case of the Midwife Toad* (New York: Random House, 1971).
5. Quoted in Koestler, *Midwife Toad*, chapter 5, 63. See also William Bateson to his wife, September 8, 1910, William Bateson Papers, Cambridge University Library; reproduced by permission of the Syndics of Cambridge University Library.
6. T. H. Morgan to Paul Maerker-Brandon, October 1, 1923, William Bateson Papers, Cambridge University Library; Raymond Pearl to Henry Donaldson, December 20, 1923, Raymond Pearl Papers, Series I. Correspondence, APS.
7. H. S. Jennings to Ross Harrison, January 21, 1924, H. S. Jennings Papers, Paul Kammerer Folder, APS.
8. See Gliboff, "Pebble and Planet," chapter 8; Peter Bowler, *The Eclipse of Darwinism* (Baltimore: Johns Hopkins University Press, 1983); Bowler, *The Non-Darwinian Revolution* (Baltimore: Johns Hopkins University Press, 1988); George Cook, "Neo-Lamarckian Experimentalism in America: Origins and Consequences," *Quarterly Review of Biology* 74 (1999): 417–437; Kathy Cooke, "The Limits of Heredity: Nature and Nurture in American Eugenics before 1915," *Journal of the History of Biology* 31 (1998): 263–278.
9. See, for example, Diane Paul, *The Politics of Heredity: Essays on Eugenics, Biomedicine and the Nature-Nurture Debate* (Albany: State University of New York Press, 1998); Robert Proctor, *Racial Hygiene: Medicine under the Nazis* (Cambridge, Mass.: Harvard University Press, 1988); Paul Weindling, *Health, Race and German Politics between National Unification and Nazism 1870–1945* [1989] (Cambridge: Cambridge University Press, 1993).
10. The most balanced accounts of the scandal are Berz and Taschwer, "Afterword"; Sander Gliboff, "The Case of Paul Kammerer: Evolution and Experimentation in the Early 20th

Century," *Journal of the History of Biology* 39 (2006): 525–563; and Hirschmüller, "Paul Kammerer."

11. Przibram, "Kammerer als Biologe"; Hans Przibram, "The Nuptial Pad of Kammerer's Water-bred *Alytes*," *Nature* 119 (1927): 635–636; Hans Przibram, "Kammerer's *Alytes*," *Nature* 118 (1926): 209–212; and Hans Przibram to Hugo Iltis, February 9, 1930. Paul Kammerer Papers, B K128, APS.
12. Ludwig Fleck, *The Genesis and Development of a Scientific Fact* (Chicago: University of Chicago Press, 1979), 173.
13. Stefan Zweig, *The World of Yesterday* [1942] (Lincoln: University of Nebraska Press, 1964), 1. On Weininger, see Chandak Sengoopta, *Otto Weininger: Sex, Science, and Self in Imperial Vienna* (Chicago: University of Chicago Press, 2000).
14. See Malachi Hacohen, "Karl Popper, the Vienna Circle, and Red Vienna," *Journal of the History of Ideas* 59 (1998): 711–734. There has been controversy about Kammerer's Jewish heritage, but his mother was Jewish; see Berz and Taschwer, "Afterword," and Eliza Slavet, "Freud's 'Lamarckism' and the Politics of Race Science," *Journal of the History of Biology* 41 (2008): 37–80.
15. On the relationship between Kammerer and Alma Mahler, see Karen Monson, *Alma Mahler, Muse to Genius: From Fin-de-siècle Vienna to Hollywood's Heyday* (New York: Houghton Mifflin, 1983). Quote from Hirschmüller, "Paul Kammerer," 33–40. For a time Mahler-Werfel worked training praying mantises in Kammerer's laboratory.
16. Bowler, *Eclipse;* Gliboff, "Pebble and Planet"; Ronald Numbers, *Darwinism Comes to America* (Cambridge, Mass.: Harvard University Press, 1998); Robert Richards, *Darwin and Evolutionary Theories of Mind and Behavior* (Chicago: University of Chicago Press, 1987); Richards, *The Meaning of Evolution* (Chicago: University of Chicago Press, 1992).
17. Gliboff, "Case of Paul Kammerer"; Sander Gliboff, *H.G. Bronn, Ernst Haeckel, and the Origins of German Darwinism: A Study in Translation and Transformation* (Cambridge, Mass.: MIT Press, 2008).
18. The experimental method, still new in the late nineteenth century, was particularly controversial in morphological development: Lynn Nyhart, *Biology Takes Form: Animal Morphology and the German Universities, 1800–1900* (Chicago: University of Chicago Press, 1995); Hans Querner, "Beobachtung oder Experiment? Die Methodenfrage in der Biologie um 1900," *Verhandlungen der deutschen zoologischen Gesellschaft* 68 (1974): 4–12. On Entwicklungsmechanik, see Jane Maienschein, "The Origins of Entwicklungsmechanik," *Developmental Biology* 7 (1991): 43–61; Wilhelm Roux, "The Problems, Methods, and Scope of Developmental Mechanics," in *Defining Biology, Lectures from the 1890s*, ed. J. Maienschein (Cambridge, Mass.: Harvard University Press, 1986), 105–148.
19. On the Vivarium, see Deborah Coen, "Living Precisely in Fin-de-Siècle Vienna," *Journal of the History of Biology* 39 (2006): 493–523; Gliboff, "Pebble and Planet," chapter 5; Gliboff, "Case of Paul Kammerer"; Veronika Hofer, "Rudolf Goldscheid, Paul Kammerer und die Biologen des Prater-Vivariums in der liberalen Volksbildung der Wiener Moderne," in *Wissenschaft, Politik und Öffentlichkeit*, ed. M. Ash and C. Stifter (Vienna: Facultas Verlags und Buchhandels AG WUV–Universitäts Verlag, 2002), 149–184; Hans Przibram, "Die neue Anstalt für experimentelle Biologie in Wien," *Verhandlungen der Gesellschaft deutscher Naturforscher und Ärzte* 74 (1902): 152–155; Hans Przibram, "Die biologische Versuchsanstalt in Wien," *Zeitschrift für biologische Technik und Methodik* 3 (1913): 163–223; and Wolfgang L. Reiter, "Zerstört und Vergessen: Die biologische Versuchsanstalt und ihre Wissenschaftler/innen," *Österreichische Zeitschrift für Geschichte* 10 (1999): 585–614.

20. Some critics viewed the products as monstrous and the Institute as too Semitic; see Gliboff, "Case of Paul Kammerer," and K. Tjebbes to William Bateson, October 1922, Bateson Correspondence, APS.
21. Mario Di Gregorio, *From Here to Eternity: Ernst Haeckel and Scientific Faith* (Göttingen: Vandenhoeck and Ruprecht, 2005); Gliboff, "Pebble and Planet"; Gliboff, *German Darwinism*; Robert Richards, *The Tragic Sense of Life: Ernst Haeckel and the Struggle over Evolutionary Thought* (Chicago: University of Chicago Press, 2008).
22. Gliboff, "Case of Paul Kammerer." Most specimens were destroyed in World War I. In the 1960s, Arthur Koestler asked Paul Weiss to duplicate the fraud with frogs. Weiss was at the Vivarium in 1926 and helped examine the injected tissue when the fraud was discovered. After attempting to duplicate it later, he concluded that injection could have been done in dead, but not in living tissue; Paul Weiss Papers, Correspondence Box 22 #8. Arthur Koestler Correspondence, RA. See also Berz and Taschwer, "Afterword"; and Gliboff, "Pebble and Planet."
23. Hirschmüller, "Paul Kammerer," 52. Both Kammerer and Tandler were prominent members of the Vienna Freemason Lodge, and Kammerer regularly presented lectures there, see Günter K. Kodek, *Unsere Bausteine sind die Menschen. Die Mitglieder der Wiener Freimaurerlogen (1869–1938)* (Vienna: Löcker, 2009). For the Freemasons' view of Kammerer, see Alfred Nimmerrichter, "Kröten und Eidechsen bestimmten sein Schicksal: Das tragische Leben des engagierten Freimaurers Paul Kammerer," in *Freimaurer, Gelehrte, Rassisten*, edited A. Nimmerrichter (Vienna: Forschungsgesellschaft Quatuor Coronati, 2007), 7–21. On Vienna's many socialist progressives, see M. G. Ash and C. H. Stifter, eds., *Wissenschaft, Politik, und Öffentlichkeit: Von der Wiener Moderene bis zur Gegenwart* (Vienna: Facultas Verlags und Buchhandels, WUV–Universitäts Verlag, 2002); Hacohen, *Karl Popper*, chapter 1; W. M. Johnston, "The Syncretist Historians of Philosophy at Vienna 1860–1930," *Journal of the History of Ideas* 32 (1971): 299–305; and Friedrich Stadler, "Spätaufklärung und Sozialdemokratie in Wien 1918–1938," in *Aufbruch und Untergang: Österreichische Kultur zwischen 1918 und 1938*, ed. F. Kadrnoska (Vienna: Europaverlag, 1981), 441–473.
24. Richard Burkhardt, *The Spirit of System: Lamarck and Evolutionary Biology* (Cambridge, Mass.: Harvard University Press, 1995). For the complex transitions in French thought, see Laurent Loison, "French Roots of French Neo-Lamarckisms, 1879–1985," *Journal of the History of Biology* 44, DOI 10.1007/s10739–010–9240-x.
25. See Richards, *Darwin, Mind, and Behavior.*
26. Ibid., 285; and Gliboff, *Origins of German Darwinism.*
27. Ernst Haeckel, *Generelle Morphologie der Organismen* (Berlin: Georg Reimer, 1866), quoted in Gliboff, "Pebble and Planet," 81; Di Gregorio, *From Here to Eternity*; and Richards, *Tragic Sense.* Daniel Gasman presents another view; see *Haeckel's Monism and the Birth of Fascist Ideology* (New York: Peter Lang, 1998). For a critique, see Robert Richards, "Ernst Haeckel's Alleged Anti-Semitism and Contributions to Nazi Biology," *Biological Theory* 2 (2007): 97–103.
28. Frederick Churchill, "August Weismann: A Developmental Evolutionist," in F. Churchill and H. Risler, eds., *August Weismann: Ausgewählte Briefe und Dokumente* (Freiburg: Universitätsbibliothek, 1999), 749–798; R. G. Winther, "August Weismann on Germ-plasm Variation," *Journal of the History of Biology* 34 (2001): 517–555. On the history of the gene concept, see P. Beurton, R. Falk, and H-J. Rheinberger, eds., *The Concept of the Gene in Development and Evolution: Historical and Epistemological Perspectives* (Cambridge: Cambridge University Press, 2000); Raphael Falk,

Genetic Analysis: A History of Genetic Thinking (Cambridge: Cambridge University Press, 2009); and Evelyn Fox Keller, *The Century of the Gene* (Cambridge, Mass.: Harvard University Press, 2002).

29. Bowler, *Eclipse*; I use the term "mechanism" in the second of two senses described by Allen, to mean mechanism as a "step-by-step efficient cause account of how the components in a system interact"; see Garland Allen, "Mechanism, Vitalism, and Organicism in Late Nineteenth and Twentieth-century Biology: The Importance of Context," *Studies in the History and Philosophy of Biology and Biomedical Sciences* 36 (2005): 261–283, 263. On purpose and progress, see Peter Bowler, *Life's Splendid Drama* (Chicago: University of Chicago Press, 1996), chapter 9; F. B. Churchill, "The Weismann-Spencer Controversy over the Inheritance of Acquired Characteristics," in *Human Implications of Scientific Advance*, ed. E. Forbes (Edinburgh: Edinburgh University Press, 1978), 451–468; Timothy Lenoir, *Strategy of Life* (Chicago: University of Chicago Press, 1982); and Richards, *Darwin, Mind, and Behavior*.
30. German/American embryologist Hans Spemann folded Pauly's organic agency into the organizer concept, for which he won the Nobel Prize; see Viktor Hamburger, Garland Allen, Jane Maienschein, and Hans Spemann, "Hans Spemann on Vitalism in Biology: Translation of a Portion of Spemann's 'Autobiography,'" *Journal of the History of Biology* 32 (1999): 231–243; and R. G. Rinard, "Neo-Lamarckism and Technique: Hans Spemann and the Development of Experimental Embryology," *Journal of the History of Biology* 21 (1988): 95–118. Loison. "French Roots," discusses the tensions between mechanism and teleology in French neo-Lamarckism.
31. Josef Breuer, "Die Krisis des Darwinismus und die Teleologie," in M. Kassowitz, *Vorträge und Besprechungen über die Krisis des Darwinismus* (Leipzig: Barth, 1902), 45–64.
32. Ibid., 49, 54, 64.
33. Gliboff, "Pebble and Planet," chapter 4; Richard Semon, *Die Mneme als erhaltendes Prinzip im Wechsel des organischen Geschehens* (Leipzig: Wilhelm Engelmann, 1908). See also Daniel Schacter, *Stranger behind the Engram: Theories of Memory and the Psychology of Science* (Mahwah, N.J.: Lawrence Erlbaum, 1982).
34. Carl Detto, *Die Theorie der direkten Anpassung und ihre Bedeutung für das Anpassungs- und Deszendenzproblem* (Jena: Fischer, 1904); Richard Semon, "Beweis für die Vererbung erworbener Eigenschaften: Ein Beitrag zur Kritik der Keimplasma Theorie," *Archiv für Rassen- und Gesellschafts-Biologie* 4 (1907): 1–46; and Semon, "Der Stand der Frage nach der Vererbung erworbener Eigenschaften," in *Fortschritte der naturwissenschaftliche Forschung* 2, ed. E. Abderhalden (Berlin: Urban und Schwarzenberg, 1911), 1–82.
35. See Ludwig Plate, "Francis Darwins Presidents Adress," *Archiv für Rassen- und Gesellschafts-Biologie* 6 (1909): 91–92.
36. August Weismann, *The Germ-Plasm Theory of Heredity* (London: Walter Schott; New York: Scribner & Sons, 1912), 408–409.
37. Paul Kammerer, "Mendelsche Regeln und Vererbung erworbener Eigenschaften," *Verhandlungen des naturforschenden Vereines in Brünn* 49 (1911): 72–110, 87, quoted by Gliboff, "Pebble and Planet," 169.
38. Daniel MacDougal to Kammerer, January 7, 1924, Desert Botanical Papers, Carnegie Institute for Science; quoted in Gliboff, "Pebble and Planet," 264.
39. Jonathan Harwood, "The Reception of Morgan's Chromosome Theory in Germany: Inter-war Debate over Cytoplasmic Inheritance," *Medizinhistorisches Journal* 19 (1984): 3–32; Jonathan Harwood, *Styles of Scientific Thought: The German Genetics Community, 1900–1933* (Chicago: University of Chicago Press, 1993). On Morgan, see Garland E. Allen, *Thomas Hunt Morgan: The Man and His Science* (Princeton: Princeton University

Press, 1978); and Robert Kohler, *Lords of the Fly: Drosophila Genetics and the Experimental Life* (Chicago: University of Chicago Press, 1994).

40. Hans Przibram, "Die Umwelt des Keimplasmas: I. Das Arbeitsprogram," *Archiv für Entwicklungsmechanik der Organismen* 33 (1911–1912): 666–681, 666.
41. Ibid., 667.
42. Paul Kammerer, "Richard Semon; Zur Wiederkehr seines Todestages," *Der Abend* 294 (1920); and Kammerer, "Mendelsche Regeln und erworbener Eigenschaften."
43. Florian Mildenberger, . . . *In der Richtung der Homosexualität verdorben. Psychiater, Kriminalpsychologen und Gerichtsmediziner über männliche Homosexualität 1850–1970* (Hamburg: Männerschwarm, 2002); Chandak Sengoopta, "Glandular Politics: Experimental Biology, Clinical Medicine, and Homosexual Emancipation in Fin-de-Siècle Central Europe," *Isis* 89 (1998): 445–473. On rejuvenation, see Chandak Sengoopta, *The Most Secret Quintessence of Life: Sex, Glands, and Hormones, 1850–1950* (Chicago: University of Chicago Press, 2006); Heiko Stoff, *Ewige Jugend. Konzept der Verjüngung vom spaten 19. Jahrhundert bis ins Dritten Reich* (Köln: Böhlau Verlag, 2003). Stoff suggests that the rejuvenation controversy destroyed Steinach's scientific reputation.
44. America was fertile ground for the rejuvenation procedure, and its effects were exaggerated in literature and film; see Laura Davidow Hirschbein. "The Glandular Solution: Sex, Masculinity, and Aging in the 1920s," *Journal of the History of Sexuality* 9 (2000): 277–304; Sengoopta, *Secret Quintessence*; D. Schultheiss, J. Denil, and U. Jonas, "Rejuvenation in the Early 20th Century," *Andrologia* 29 (1997): 351–355; and Heiko Stoff, "Vermännlichung und Verweiblichung: Wissenschaftliche und utopische Experimente im frühen 20. Jahrhundert," in *Wahrnehmung und Herstellung von Geschlecht*, ed. Ursula Pasero and Friederike Braun (Opladen: Westdeutscher Verlag, 1999), 47–62.
45. The Eugen Steinach–Harry Benjamin Correspondence, Historical Collections, NYAM. In his remembrance—Harry Benjamin, "Eugen Steinach, 1861–1944: A Life of Research," *Scientific Monthly* 61 (1945): 427–442—Benjamin compared Steinach's impact to those of Freud and Darwin. On Benjamin, see Leah C. Schaefer and Connie C. Wheeler, "Harry Benjamin's First Ten Cases (1938–1953): A Clinical Historical Note," *Archives of Sexual Behavior* 24 (1995): 73–93.
46. Kammerer to Benjamin, March 10, 1922, Harry Benjamin Collection, Correspondence, Paul Kammerer Folder 1, KI.
47. Benjamin to Kammerer, December 8, 1922, emphasis in original. Steinach was opposed to Kammerer's American lecture tour; see Kammerer to Benjamin, June 18, 1922, Harry Benjamin Collection, Correspondence, Paul Kammerer Folder 1, KI.
48. Steinach, *Sex and Life*, xvi.
49. Kammerer quoting Steinach, Kammerer to Glücksmann, August 19, 1920. Documents Collection, Kammerer Correspondence, ÖNB.
50. The Vivarium was then owned by the Austrian Academy of Sciences. Annual reports from the early 1920s show its desperate financial condition and stress the importance of money for Steinach's research. Vivarium Papers, ÖAW.
51. Steinach to Benjamin, October 23, 1923, emphasis in original. The Eugen Steinach–Harry Benjamin Correspondence, Historical Collections, NYAM.
52. *New York Times*, June 3, 1923. The misstatement was traced to endocrinologist Eduard Uhlenhuth, an associate at the Rockefeller Institute. From 1925 to 1955, Uhlenhuth was professor and then dean at the University of Maryland School of Medicine. He had worked at the Vivarium and probably knew both Kammerer and Steinach.
53. Kammerer to Benjamin, June 29, 1923. Harry Benjamin Collection, Correspondence, Paul Kammerer Folder 2, KI.

54. Kammerer to Steinach, July 5, 1923, emphasis in original. Harry Benjamin Collection, Correspondence, Paul Kammerer Folder 2, KI. The folder contains two letters with different endings. One has the quoted section as the last paragraph; the second does not. That there were two suggests Kammerer's trepidation in confronting Steinach.
55. Steinach also praised Kammerer's 1919 review of his work. After the *Times* blowup, Kammerer was surprised that Steinach still agreed to write the forward for his book on rejuvenation (Paul Kammerer, *Rejuvenation and the Prolongation of Human Efficiency: Experiments with the Steinach-operation on Man and Animals* (New York: Boni and Liveright, 1923)), Harry Benjamin Collection, Correspondence, Paul Kammerer Folder 2, KI.
56. Printed advertisement for Kammerer's *Rejuvenation* book, Eugen Steinach–Harry Benjamin Correspondence, Historical Collections, NYAM.

4. SEX, RACE, AND HEAT RATS

Portions of this chapter are revised from Cheryl Logan, "Overheated Rats, Race, and the Double Gland: Paul Kammerer, Endocrinology, and the Problem of Somatic Induction," *Journal of the History of Biology* 40 (2007): 683–725.

1. The word "race" is not intended to reflect biological reality. Rather, I use it to reflect a time when the existence of fixed racial hierarchies was a common scientific and social assumption.
2. Archives of the University of Vienna, Paul Kammerer Papers, folder 947, 1907–1908, ÖAW.
3. See also Paul Kammerer, *Ursprung der Geschlechtsunterschiede* (Berlin: Urban and Schwarzenberg, 1912).
4. Ibid., 40.
5. Paul Kammerer, "Steinachs Forschungen über Entwicklung, Beherrschung und Wandlung der Pubertät," *Ergebnisse der inneren Medizin und Kinderheilkunde* 17 (1919): 295–398.
6. Ibid., 348.
7. The topic was of great interest; see Steven Brush, "Nettie M. Stevens and the Discovery of Sex Determination by Chromosomes," *Isis* 69 (1978): 162–172; Sharon Kingsland, "Maintaining Continuity through a Scientific Revolution: A Rereading of E. B. Wilson and T. H. Morgan on Sex Determination and Mendelism," *Isis* 98 (2007): 468–488; Jane Maienschein, "What Determines Sex? A Study of Converging Approaches, 1880–1916," *Isis* 75 (1984): 456–480; Susan Mosedale, "Science Corrupted: Victorian Biologists Consider 'The Woman Question,'" *Journal of the History of Biology* 11 (1978): 1–55; Marsha L. Richmond, "Richard Goldschmidt and Sex Determination: The Growth of German Genetics, 1900–1935," Ph.D. dissertation, Indiana University, 1986.
8. Kammerer, "Steinachs Forschungen," 363.
9. Ibid., 364. Kammerer also argued that the link was evident in the cyclic relationship between ovulation and the production of hormones by the corpus luteum, which was full of interstitial cells.
10. Eugen Steinach and Paul Kammerer, "Klima und Mannbarkeit," *Archiv für Entwicklungsmechanik* 46 (1920): 391–458.
11. Ibid., 401.
12. Presumably because the males were more like the control females, Steinach and Kammerer described this as feminization.
13. Ibid., 404.
14. Ibid., 442.

15. On Steinach and rats in experimental physiology, see Cheryl Logan, "'[A]re Norway rats . . . things?': Diversity versus Generality in the Use of Albino Rats in Experiments on Development and Sexuality," *Journal History of Biology* 34 (2001): 287–314.
16. The German flag was hoisted on the island of Matupi, part of the Bismarck Archipelago, in November 1884.
17. Reche later cofounded the German Society for Blood Group Research and the *Zeitschrift für Rassenphysiologie*, and he warned of the negative impact of intermixing "alien" blood groups; see Robert Proctor, *Racial Hygiene: Medicine under the Nazis* (Cambridge, Mass.: Harvard University Press, 1988), chapter 6. On Reche's career, see Katje Geisenhainer, "*Rasse ist Schicksal*": *Otto Reche (1879–1966)— ein Leben als Anthropologe und Völkerkundler* (Leipzig: Evangelische Verlagsanstalt, 2002); and Thomas Mayer, "Familie, Rasse, und Genetik: Deutschnationale Eugenikern im Österreich in der Zwischenkriegszeit," in G. Baader, V. Hofer, and T. Mayer, eds., *Eugenik in Österreich:. Biopolitische Strukturen von 1900–1945* (Vienna: Czernin, 2007), 162–183.
18. Steinach and Kammerer, "Klima," 414.
19. Ibid., 420.
20. Ibid., 416.
21. Ibid., 427.
22. Sander Gilman, *Difference and Pathology: Stereotypes of Sexuality, Race and Madness* (Ithaca: Cornell University Press, 1985); Pascal Grosse, *Kolonialismus, Eugenik und Bürgerliche Gesellschaft in Deutschland, 1850–1918* (Frankfurt: Campus, 2000); Andrew Zimmerman, *Anthropology and Antihumanism in Imperial Germany* (Chicago: University of Chicago Press, 2001). On differential birth rates between "primitive" peoples and Europeans, see Brian Dolan, ed., *Malthus, Medicine & Morality: Malthusianism after 1798* (Amsterdam: Rodopi, 2000).
23. Steinach and Kammerer, "Klima," 437.
24. Steinach and Kammerer extended the implications of the double gland to the question of premenstrual sexual intercourse (see "Klima," 437–439).
25. Ibid., 427. Their explanation relied on psychophysical adaptation: increases in heat would have greater impact in cultures in which people had already adjusted to colder climates than in those accustomed to warmer ones.
26. Ibid., 423, 444.
27. Ibid., 446.
28. Ibid., 446.
29. Kammerer so stated in *Rejuvenation*, xv. He also told Benjamin that Steinach did not understand the implications of their work; see Kammerer to Benjamin, March 10, 1922, Harry Benjamin Papers, Correspondence, Paul Kammerer Folder 1, KI.
30. For the debates in anthropology at the end of the nineteenth century and their relation to race and evolution, see Geisenhainer, "*Rasse ist Schicksal*"; Benoit Massin, "From Virchow to Fischer: Physical Anthropology and 'Modern Race Theories' in Wilhelmine Germany," in *Volksgeist as Method and Ethic: Essays on Boasian Ethnography and the German Anthropological Tradition*, ed. G. W. Stocking, Jr. (Madison: University of Wisconsin Press, 1996), 79–154; and Zimmerman, *Anthropology and Antihumanism*.
31. Geisenhainer, "*Rasse ist Schicksal*," 54.
32. As they criticized predeterministic explanations, Steinach and Kammerer also sometimes followed them, using hierarchical terms. For example, they did not challenge the idea that women and Naturvölker were infantile compared to men.
33. Massin, "From Virchow to Fischer," 125. On Austrian anthropology and neo-Lamarckism, see Doris Byer, "Evolutionistische Anthropologen: Zur Ambivalenz eines

hundertjährigen Fortschrittsparadigmas," in M. Ash and C. Stifter, eds., *Wissenschaft, Politik und Öffentlichkeit: Von der Wiener Moderne bis zur Gegenwart* (Vienna: Facultas Verlags und Buchhandels AG WUV–Universitäts Verlag, 2002), 185–206.

34. Kammerer's Viennese contemporary Robert Stigler took a radically different approach from Steinach and Kammerer's animal experiments. Promising them medicine for their children if they would talk about their sex lives, Stigler questioned African women about their sex drives and sexual satisfaction; Robert Stigler, "Vergleichende Sexualphysiologie der schwarzen und der weißen Frau," *Verhandlungen des I. Internationalen Kongreses für Sexualforschung, Zweiter Band, Physiologie, Pathologie und Therapie* (Berlin: Marcuse & Weber, 1926). Stigler's measures of "race physiology" consisted of descriptions of coital posture, masturbation, etc., and he concluded that anyone who searched for sexual motives in females was in error.
35. See Jonathan Harwood, *Styles of Scientific Thought: The German Genetics Community, 1900–1933* (Chicago: University of Chicago Press, 1993); Richmond, "Sex Determination"; and Jan Sapp, *Beyond the Gene: Cytoplasmic Inheritance and the Struggle for Authority in Genetics* (Oxford: Oxford University Press, 1987).
36. Artur Biedl, *Innere Sekretion*, vol. 2 (Berlin: Urban und Schwarzenberg, 1916). Footnotes indicate that Biedl and Kammerer discussed Steinach's ideas.
37. American embryologist Carl Moore challenged Steinach's view. See Chandak Sengoopta, *The Most Secret Quintessence of Life: Sex, Glands, and Hormones, 1850–1950* (Chicago: University of Chicago Press, 2006), chapter 4; and Per Södersten, "A Historical and Personal Perspective on the Aromatization Revolution: Steinach Confirmed," in *Brain Aromatase, Estrogens, and Behavior*, ed. Jacques Balthazart and Gregory Ball (Oxford: Oxford University Press, 2012).
38. Alexander Lipschütz, *The Internal Secretions of the Sex Glands* (Cambridge: Heffer & Sons, 1924), 120–121.
39. Kolmer died suddenly in 1931. In 1934, Arnold Durig, then director of the Institute, nominated Steinach for the Nobel Prize, his sixth of seven nominations.
40. Yoshitaka Kitahara, "Über die Entstehung der Zwischenzellen der Keimdrüsen des Menschen und der Säugetiere und über deren physiologische Bedeutung," *Archiv für Entwicklungsmechanik der Organismen* 52 (1923): 550–615.
41. Udo Schagen, "Die Forschung an menschlichen Organen nach 'plötzlichem Tod' und der Anatom Hermann Stieve (1886–1952)," in *Die Berliner Universität in der NS-Zeit Bd. II: Fachbereiche und Fakultäten*, ed. Rüdiger vom Bruch (Stuttgart: Franz Steiner Verlag, 2005), 35–54. Details of Stieve's life and work are described in B. Romeis, "Hermann Stieve," *Anatomischer Anzeiger* 99 (1952): 401–440.
42. Hermann Stieve, "Entwicklung, Bau, und Bedeutung der Keimdrüsenzwischenzellen," *Ergebnisse der Anatomie und Entwicklungsgeschichte* 23 (1921): 1–249.
43. Ibid., 57.
44. Knud Sand, "Experiments on the Endocrinology of the Sexual Glands," *Endocrinology* 7 (1923): 273–301, 300. Sand addressed the issue of histological origin, which did not alter his conclusion.
45. Alexander Lipschütz, "New Experimental Data on the Question of the Seat of the Endocrine Function of the Testicle," *Endocrinology* 7 (1923): 1–18.
46. Jürgen Harms, *Körper und Keimzellen* (Berlin: Springer, 1926). The work merits a comprehensive study in the history of endocrinology and sexuality.
47. Scientific evidence was complicated by the fact that different authors proposed different types of interstitial cells. Some considered fetal interstitial cells different from adult interstitial cells.

48. Harms, *Körper*, 771.
49. Ibid., 100, 105, 99.
50. Ibid. In Harms's view, the effects of alcohol and radiation were direct, while those of hunger, illness, overfeeding, and climate were indirect.
51. Ibid., 762. Stieve housed adult mice in very high heat (37 degrees Celsius). Many animals died, but a few survived and showed evidence of sex drives. He attributed the changes not to temperature, but to metabolic transformations of the whole body.
52. Ibid., 765.
53. Stieve, "Bedeutung der Keimdrüsenzwischenzellen," 110.
54. See Nelly Oudshoorn, "Endocrinologists and the Conceptualization of Sex, 1920–1940," *Journal of the History of Biology* 23 (1990): 42–83, where Oudshoorn develops this relativism for a later period.
55. Hermann Stieve, *Unfruchtbarkeit als Folge unnatürlicher Lebensweise: Ein Versuch, die ungewollte Kinderlosigkeit des Menschen auf Grund von Tierversuchen und anatomischen Untersuchungen auf die Folgen des Kulturlebens zurückzuführen* (Munich: Bergmann, 1926), 40, 47.
56. Ibid., 61. Stieve was probably just as uncomfortable with Kammerer's view of race. He wrote that Jews could be considered a "*Mastrasse*"—a race so fat and inactive that it had become almost infertile.
57. Felix Sieglbauer, "Zur Frage der Zwischenzellen," *Archiv für mikroskopische Anatomie und Entwicklungsgeschichte* 100 (1924): 473–487. For the cells' application in the United States, see Max Thorek. "Experimental Investigations of the Role of the Leydig, Seminiferous, and Sertoli Cells and Effects of Testicular Transplants," *Endocrinology* 8 (1924): 61–90.
58. Sieglbauer, "Frage der Zwischenzellen," 479, 481.
59. Sengoopta, *Secret Quintessence*; and Heiko Stoff, *Ewige Jugend: Konzept der Verjüngung vom spaten 19. Jahrhundert bis ins Dritten Reich* (Köln: Böhlau Verlag, 2003). On birth control and sexual reform in the interwar period, see Atina Grossmann, *Reforming Sex: The German Movement for Birth Control and Abortion Reform, 1920–1950* (Oxford: Oxford University Press, 1995).
60. Letter from Alexander Lipschütz to George Corner, November 22, 1963, George W. Corner Papers, APS. I thank Roy Goodman for suggesting Corner's significance.
61. In his posthumous tribute to Steinach, Benjamin described a meeting in which several histologists accepted the secretory hypothesis; Harry Benjamin, "Eugen Steinach, 1861–1944: A Life of Research," *Scientific Monthly* 61 (1945): 427–442. The interstitial cells were finally confirmed as a central source of mammalian sex steroids in the 1960s.

5. "PRODUCTIVE" EUGENICS

1. Paul Kammerer, *Sind wir Sklaven der Vergangenheit oder Werkmeister der Zukunft? Anpassung, Vererbung, Rassenhygiene in dualistischer and monistischer Betrachtungsweise* (Vienna: Anzengruber-Verlag, 1913).
2. Paul Kammerer, *Lebensbeherrschung. Grundsteinlegung zur organischen Technik, 2nd ed.* (Hamburg: Paul Hartung, 1920), 4.
3. See *Archiv für Rassen- und Gesellschafts-Biologie* 6.2 (1909): 10. Leonard Darwin was president of the First International Congress of Eugenics.
4. For a sampling of the literature in the history of eugenics, see Mark Adams, ed., *The Wellborn Science* (New York: Oxford University Press, 1990); Gerhardt Baader, Veronika

Hofer, and Thomas Mayer, eds., *Eugenik in Österreich: Biopolitische Strukturen von 1900–1945* (Vienna: Czernin, 2007); Ian Dowbiggin, *The Sterilization Movement and Global Fertility in the Twentieth Century* (New York: Oxford University Press, 2008); Daniel Kevles, *In the Name of Eugenics* [1985] (Cambridge, Mass.: Harvard University Press, 1995); Wendy Kline, *Building a Better Race* (Berkeley: University California Press, 2005); Pauline Mazumdar, *Eugenics, Human Genetics, and Human Failings* (New York: Routledge, 1992); Diane Paul, *The Politics of Heredity: Essays on Eugenics, Biomedicine and the Nature-Nurture Debate* (Albany: State University of New York Press, 1998); Nancy Stepan, *The Hour of Eugenics: Race, Gender and Nation in Latin America* (Ithaca: Cornell University Press, 1996); Richard Soloway, *Demography and Degeneration: Eugenics and the Declining Birthrate in Twentieth Century Britain* (Chapel Hill: University of North Carolina Press, 1995); Alexandra Stern, *Eugenic Nation* (Berkeley: University of California Press, 2005); Marius Turdus, *Modernism and Eugenics (*Houndmills, U.K.: Palgrave Macmillan, 2010); Paul Weindling, *Health, Race and German Politics between National Unification and Nazism 1870–1945* [1989] (Cambridge: Cambridge University Press, 1993); Peter Weingart, Jürgen Kroll, and Kurt Bayertz, *Rasse, Blut und Gene* (Frankfurt am Main: Surkampf, 1992); and Sheila Weiss, *Race, Hygiene and National Efficiency: The Eugenics of Wilhelm Schallmayer* (Berkeley: University of California Press, 1987). On connections between American and Nazi eugenics, see Stefan Kühl, *The Nazi Connection: Eugenics, American Racism, and German National Socialism* (New York: Oxford University Press, 1994); Jonathan Spiro, *Defending the Master Race: Conservation, Eugenics, and the Legacy of Madison Grant* (Lebanon, N.H.: University Press of New England, 2009).

5. Lewellys Barker, "Foreword," in M. Aldrich, W. Carruth, and C. Davenport, eds., *Eugenics, Twelve University Lectures* (New York: Dodd and Mead, 1914), x, emphasis in original.
6. Peter Bowler, "Holding Your Head up High: Degeneration and Orthogenesis in Theories of Human Evolution," in *History, Humanity and Evolution*, ed. J. R. Moore (Cambridge: Cambridge University Press, 1989), 329–353; Frederick B. Churchill, "Living with the Biogenetic Law: A Reappraisal," in *From Embryology to Evo-Devo: A History of Developmental Evolution*, ed. Manfred D. Laubichler and Jane Maienschein (Cambridge, Mass.: MIT Press, 2007), 37–81; Stephen J. Gould, *Ontogeny and Phylogeny* (Cambridge, Mass.: Harvard University Press, 1977); Laura Otis, *Organic Memory: History and the Body in the Late Nineteenth and Early Twentieth Centuries* (Lincoln: University of Nebraska Press, 1994); and Robert Richards, *The Meaning of Evolution* (Chicago: University of Chicago Press, 1992). For a discussion of social context and class, see Paul White, "Acquired Character: The Hereditary Materials of the 'Self-made Man,'" in S. Müller-Wille and H. J. Rheinberger, eds., *Heredity Produced: At the Crossroads of Biology, Politics, and Culture, 1500–1870* (Cambridge, Mass.: MIT Press, 2007), 375–397.
7. Arthur Estabrook, *The Jukes in 1915* (Washington, D.C.: Carnegie Institution of Washington, 1916), iii. On Dugdale, see Elof Carlson, "R. L. Dugdale and the Jukes Family: A Historical Injustice Corrected," *Biosciences* 30 (1980): 535–539. Estabrook noticed much change in the new environments; but his boss, Davenport, altered the published story; see Nathaniel Deutsch, *Inventing America's Worst Family* (Berkeley: University of California Press, 2009).
8. Paul Kammerer, *Neuvererbung oder Vererbung erworbener Eigenschaften: Erbliche Belastung und erbliche Entlastung* (Stuttgart: Walter Seifert, 1925). See also Sander Gliboff, "'Protoplasm is soft wax in our hands': Paul Kammerer and the Art of Biological Transformation," *Endeavour* 29 (2005): 162–167. Kammerer refuted critics who interpreted his zoological demonstrations as degenerations to a primitive evolutionary state.

9. James Collins, Scott Gilbert, Manfred Laubichler, and Gerd Müller, "Modeling in EvoDevo: How to Integrate Development, Evolution, and Ecology," in Manfred Laubichler and Gerd Müller, eds., *Modeling Biology: Structures, Behavior, Evolution* (Cambridge, Mass.: MIT Press, 2007), 355–378; Hans Przibram, "Die neue Anstalt für experimentelle Biologie in Wien," *Verhandlungen der Gesellschaft deutscher Naturforscher und Ärzte* 74 (1902): 152–155. Stephanie Schmidt in "Homeosis and Atavistic Regeneration: The 'Biogenetic Law' in Entwicklungsmechanik," *History and Philosophy of the Life Sciences* 25 (2003): 193–210, stresses the compatibility between Wilhelm Roux's developmental mechanics and interpretations of development that drew on the biogenetic law. Roux's journal, *Archiv für Entwicklungsmechanik*, published much Vivarium research, including Kammerer's; see S. J. Counce, "Archives for Developmental Mechanics: W. Roux, Editor (1894–1924)," *Roux's Archive for Developmental Biology* 204 (1994): 79–92.
10. The fate of "euthenics" reflects the absence. American reformers briefly used the term to refer to human biological improvement via environmental impact, but it disappeared; see Ellen Swallow Richards, *Euthenics, the Science of Controllable Environment: A Plea for Better Living Conditions as a First Step toward Higher Human Efficiency* (Boston: Whitcomb and Borrows, 1910).
11. Paul Kammerer, *The Inheritance of Acquired Characteristics* (New York: Boni and Liveright, 1924), and Kammerer, *Neuvererbung*. Though the two works cover much the same material, one is not just a translation of the other. The English version is based on Kammerer's American lectures.
12. Kammerer, *Inheritance of Acquired Characteristics*, 354.
13. Ibid., 354–355. Kammerer noted regular press criticism of his challenges to race hygiene during his American lecture tour.
14. Peter Bowler, "E. W. MacBride's Lamarckian Eugenics and Its Implications for the Social Construction of Scientific Knowledge," *Annals of Science* 41 (1984): 245–260. Many Lamarckians did not reject racial approaches to eugenics; see Stepan, *Hour of Eugenics;* Nancy Stepan, "Eugenics in Brazil, 1917–1940," in *The Wellborn Science*, ed. M. Adams (New York: Oxford University Press, 1990), 110–152; and Charlotte Sleigh, "The Ninth Mortal Sin: The Lamarckism of W. M. Wheeler," in A. Lustig, R. Richards, and M. Ruse, eds., *Darwinian Heresies* (Cambridge: Cambridge University Press, 2004), 151–172.
15. See Paul Kammerer, "Allgemeine Symbiose und Kampf ums Dasein als gleichberechtigte Triebkräfte der Evolution," *Archiv für Rassen- und Gesellschafts-Biologie* 6 (1909): 585–608; and Kammerer, "Kampf ums Dasein und gegenseitige Hilfe, part III." *Einzeltod, Völkertod, biologische Unsterblichkeit* (Vienna: Anzengruber Verlag, 1918). Sleigh, "Ninth Mortal Sin," describes Wheeler's Lamarckian use of cooperation.
16. Kammerer, *Inheritance of Acquired Characteristics*, 262.
17. Charles Darwin, *The Descent of Man and Selection in Relation to Sex, vol. 1,* 1st ed. (New York: Appleton, 1872), chapter 3. Kammerer, *Einzeltod, Völkertod, biologische Unsterblichkeit*, 12 and 14. Kammerer also saw cell division as a relationship of exchange and protection.
18. Paul Kammerer, "Vererbung erzwungener Fortpflanzungsanpassungen I. und II. Mitteilung: Die Nachkommen der spätgeborenen *Salamandra maculosa* und der frühgeborenen *Salamandra atra*," *Archiv für Entwicklungsmechanik* 25 (1917): 7–51. Kammerer's comparison is very close to the modern distinction between r and K selection; see L. Partridge and P. H. Harvey, "The Ecological Context of Life History Evolution," *Science* 241 (1988): 1449–1455.
19. Kammerer, *Einzeltod, Völkertod, biologische Unsterblichkeit*, 59.

20. Ibid., 63.
21. Paul Kammerer, *Das biologische Zeitalter: Fortschritte der organischen Technik* (Vienna: Verein Freie Schule, 1920), 6.
22. Paul Kammerer, *Rejuvenation and the Prolongation of Human Efficiency: Experiments with the Steinach-operation on Man and Animals* (New York: Boni and Liveright, 1923).
23. Ibid., 25.
24. Ibid., 36. The endocrine system often provided physiological justification for a holism that challenged reductionistic physiology; see Sarah Tracy, "George Draper and American Constitutional Medicine, 1916–1946: Reinventing the Sick Man," *Bulletin History of the Medicine* 66 (1992): 53–89. I thank Tom Rosenbaum for directing me to this source.
25. Kammerer, *Rejuvenation,* 46.
26. Ibid., 28.
27. Ibid., 26.
28. Kammerer, *Biologisches Zeitalter,* 16, 12.
29. Ibid., 13. "Audacious" organotherapy refers to the ridiculed efforts of Charles Brown-Sequard, who in 1889 maintained that he had been invigorated by consuming the ground testicles of animals; see Merriley Borrell, "Organotherapy, British Physiology, and Discovery of the Internal Secretions," *Journal of the History of Biology* 9 (1978): 235–268.
30. Kammerer, *Inheritance of Acquired Characteristics,* 263 and 264. See also Kammerer, *Biologisches Zeitalter* and *Einzeltod, Völkertod, biologische Unsterblichkeit.*
31. Kammerer, *Inheritance of Acquired Characteristics,* 265.
32. Kammerer, *Rejuvenation,* 39.
33. Hugo Iltis, "Der Kampf um Darwinismus," *Arbeiter-Jahrbuch (*Bodenbach: Gärtner, 1926). Iltis called Darwin "our gallant comrade (*Bundesgenosse*)" in the struggle against clericalism, challenging Houston Stewart Chamberlain, the British-German racist who considered Darwinism "a foreign growth grafted on to the German mind" (2). Sheila Weiss suggests that several eugenicists developed views compatible with socialism. But threats about the "Slavization of Germany" produced a shift to race; see Weiss, "The Race Hygiene Movement in Germany," *Osiris* 3 (1986): 212; and Weiss, *Race, Hygiene and National Efficiency.* See also Michael Schwartz, *Sozialistische Eugenik: Eugenische Sozialtechnologien in Debatten und Politik der deutschen Sozialdemokratie, 1890–1933* (Bonn: J.H.W. Dietz, 1995); and Alberto Spektorowski, "The Eugenic Temptation in Socialism: Sweden, Germany, and the Soviet Union," *Comparative Studies of Society and History* 46 (2004): 84–106.
34. Iltis used the term "direct" to mean experimental evidence that could be sensed immediately, not inferred from the past, as was done in paleontology and developmental morphology.
35. Iltis, "Kampf um Darwinismus," 7, 8, and 9. Darwinism in the broad sense also supported the difference between communism and socialism. This was important for Austrian socialists, who sought a blend of socialism and democracy. Iltis eventually elaborated his critique in several works challenging racism; see Eliza Slavet, "Freud's 'Lamarckism' and the Politics of Race Science," *Journal of the History of Biology* 41 (2008): 37–80, and Paul Weindling. "Central Europe Confronts German Racial Hygiene: Friedrich Hertz, Hugo Iltis, and Ignaz Zollschan as Critics of Racial Hygiene," in M. Turda and P. Weindling, eds., *"Blood and Homeland": Eugenics and Racial Nationalism in Central and Southeast Europe, 1900–1940* (Budapest: Central European University Press, 2007), 263–280.

36. Garland Allen, *Thomas Hunt Morgan: The Man and His Science* (Princeton: Princeton University Press, 1978). Morgan's theory was published in book form in 1915; in 1933, he won the Nobel Prize for his work in genetics.
37. See, for example, L. Sofer, "Über die Plastizität der menschlichen Rassen," *Archiv für Rassen- und Gesellschafts-Biologie* 5 (1908): 660–668.
38. G. Baader, "Eugenische Programme in der sozialistischen Parteienlandschaft in Deutschland und in Österreich im Vergleich," in G. Baader, V. Hofer, and T. Mayer, eds., *Eugenik in Österreich: Biopolitische Strukturen von 1900–1945* (Vienna: Czernin, 2007), 66–139; and Michael Schwartz, "'Proletarier' und 'Lumpen': Sozialistische Ursprünge eugenischen Denkens," *Vierteljahrshefte für Zeitgeschichte* 42 (1994): 537–570.
39. Kammerer. *Lebensbeherrschung*, 16.
40. Ibid., 17.
41. Ibid.
42. Ibid., 14, emphasis added.
43. Paul Kammerer. "Kultur und Rasse," *Einzeltod, Völkertod, biologische Unsterblichkeit*, part 2 (Vienna: AnzengruberVerlag, 1918), 9. On Hertz, see Weindling, "Central Europe Confronts Racial Hygiene." Hertz's work was then titled *Rasse und Kultur: Eine kritische Untersuchung der Rassentheorien* (Leipzig: Alfred Kröner, 1915).
44. Correspondence between Kammerer and Davenport, December 17, 1923–January 31, 1924. Charles B. Davenport Papers, Series I, Box 60, APS. Davenport probably never tried the experiments, but the visit occurred in 1924.
45. Paul Kammerer, "Ist die Rasse veränderlich?" *Der Morgen* 2 (1926): 323–338, 332, 333.
46. Paul Kammerer, *Das Ratzel der Vererbung: Grundlagen der allgemeine Vererbungslehre* (Berlin: Ullstein, 1925), 129. I thank Klaus Taschwer for directing me to this and several other sources.
47. Kammerer, *Lebensbeherrschung*, 17 and 18.
48. Kammerer, *Biologisches Zeitalter*, 17.
49. Ibid., 18.
50. Chemist Wilhelm Ostwald, neurologist August Forel, and biologist Ludwig Plate followed Haeckel as founders of monism. They often disagreed with Haeckel, and their debates led to many varieties of monism; see P. Ziche, ed., *Monismus um 1900: Wissenschaftskultur und Weltanschaaung* (Berlin: Wissenschaft and Bildung, 2000); Weindling, *Health, Race and German Politics*, 131–133. Kammerer's importance is shown by his presence as one of six contributors to the Festschrift celebrating Wilhelm Ostwald's sixtieth birthday.
51. For contrasting views of Haeckel, see Daniel Gasman, *Haeckel's Monism and the Birth of Fascist Ideology* (New York: Peter Lang, 1998); Sander Gliboff, "The Pebble and the Planet: Paul Kammerer, Ernst Haeckel, and the Meaning of Darwinism," Ph.D. dissertation, Johns Hopkins University, 2001; Gliboff, *H. G. Bronn, Ernst Haeckel, and the Origins of German Darwinism* (Cambridge, Mass.: MIT Press, 2008); Veronika Hofer, "Rudolf Goldscheid, Paul Kammerer und die Biologen des Prater-Vivariums in der liberalen Volksbildung der Wiener Moderne," in M. Ash and C. Stifter, eds., *Wissenschaft, Politik und Öffentlichkeit* (Vienna: Facultas Verlags und Buchhandels AG WUV–Universitäts Verlag, 2002), 149–184; Robert Richards, *The Tragic Sense of Life: Ernst Haeckel and the Struggle over Evolutionary Thought* (Chicago: University of Chicago Press, 2008); Heiko Weber, "Der Monismus als Theorie einer einheitlichen Weltanschaaung am Beispiel der Positionen von Ernst Haeckel und August Forel," in *Monismus um 1900: Wissenschaftskultur und Weltanschaaung*, ed. P. Ziche (Berlin: Verlag für Wissenschaft und Bildung, 2000), 81–127. Richards, Gliboff, and Weikart refute accounts that interpret

monism as a mystical movement leading directly to Hitler; see Robert Richards, "Ernst Haeckel's Alleged Anti-Semitism and Contributions to Nazi Biology," *Biological Theory* 2 (2007): 97–103. On the distinctions between the Austrian and German monist movements, see Richard Weikart, *Socialist Darwinism, Evolution in German Socialist Thought from Marx to Bernstein* (San Francisco: International Scholars Publications, 1999), and "'Evolutionäre Aufklärung'? Zur Geschichte des Monistenbundes," in Ash and Stifter, eds., *Wissenschaft, Politik und Öffentlichkeit* (Vienna: Facultas Verlags und Buchhandels AG WUV–Universitäts Verlag, 2002), 137–148.

52. On monism and the Vienna Circle, see Friedrich Stadler, "Spätaufklärung und Sozialdemokratie in Wien, 1918–1938," in *Aufbruch und Untergang: Österreichische Kultur Zwischen 1918 und 1938*, ed. F. Kadrnoska (Vienna: Europaverlag, 1981), 441–473, 458.

53. Paul Kammerer, "Dualismus in der Vererbungswissenschaft," *Die monistische Monatshefte* 11 (1926): 18–23. Monism's commitment to the inheritance of acquired characteristics is also clear in the connection between Forel and Semon. Forel rejected racial approaches to hygiene and with Semon founded an International Order for Ethics and Culture; see Weber, "Monismus als einheitlichen Weltanschaaung"; and Weindling, *Health, Race and German Politics.* Forel's student, psychiatrist Adolf Meyer, and physiologist Jacques Loeb were among those who brought monism to America.

54. Paul Kammerer, "Kampf ums Dasein und gegenseitige Hilfe," part 3, in *Einzeltod, Völkertod, biologische Unsterblichkeit* (Vienna: Anzengruber-Verlag, 1918), 39 and 40. Austrian monism stressed disseminating science to the lay public, and Kammerer repeatedly took part in public education programs; see Hofer, "Biologen des Prater-Vivariums."

55. Kammerer, *Biologisches Zeitalter*, 18.

56. See Weindling, "Central Europe Confronts Racial Hygiene."

57. In 1894, Jodl founded the Vienna Ethical Society; see William Johnston. "Syncretist Historians of Philosophy at Vienna 1860–1930," *Journal of the History of Ideas* 32 (1971): 299–305.

58. Friedrich Jodl, "Der Monismus und die Kulturprobleme der Gegenwart," in *Die erste internationale Monisten-Kongreß in Hamburg vom 8–11 September 1911*, ed. W. Bloßfeld (Leipzig: Alfred Kröner Verlag, 1912), 117–130, 121 and 122.

59. Ibid., 123.

60. Weber, "Monismus als einheitlichen Weltanschaaung."

61. Letter from Paul Kammerer to Margarete Jodl, November 1, 1914. Wilhelm Börner Collection, Correspondence, WBR.

62. Kammerer, *Neuvererbung*, 81 and 179, from the Vienna City Library (WBR). The book contains the stamp "Library Wilhelm Börner."

63. "Zur Erinnerung an Paul Kammerer," *Die Neue Freie Presse*, September 26, 1936 (No. 25878).

64. On Goldscheid, see Gudrun Exner, "Rudolf Goldscheid (1870–1931) and the Economy of Human Beings," *Vienna Yearbook of Population Research* 2 (2004): 283–301; Wolfgang Fritz and Gertraude Mikl-Horke, *Rudolf Goldscheid—Finanzsoziologie und ethische Sozialwissenschaft* (Vienna: Lit Verlag, 2007); Gertraude Mikl-Horke, "Max Weber und Rudolf Goldscheid: Kontrahenten in der Wendezeit der Soziologie," *Sociologia Internationalis* 42 (2004): 265–286; Gertraude Mikl-Horke, "An Old Idea of 'Human Economy' and the New Global Finance Capitalism," *sociology the european electronic newsletter* 7 (2005): 36–42; and Georg Witrisal, "Der 'Soziallamarckismus' Rudolf Goldscheids: Ein milieutheoretischer Denker zwischen humanitärem Engagement und Sozialdarwinismus," Master's thesis, University of Graz, 2004. On Kammerer and Goldscheid, see Hofer, "Biologen des Prater-Vivariums"; and Jochen Fleischhacker, "Menschen—und

Güterökonomie—Anmerkungen zu Rudolf Goldscheids demoökonomischen Gesellschaftsentwurf," in M. Ash and C. Stifter, eds., *Wissenschaft, Politik und Öffentlichkeit* (Vienna: FacultasVerlags und Buchhandels AG WUV–Universitäts Verlag, 2002), 207–229.

65. Spektorowski, "Eugenic Temptation in Socialism,"85.
66. Rudolf Goldscheid, *Höherentwicklung und Menschenökonomie: Grundlegung der Sozialbiologie* (Leipzig: Klinkhardt, 1911), 528, quoted in Exner, "Rudolf Goldscheid," 289.
67. Fritz and Mikl-Horke, *Rudolf Goldscheid*, 138 and 157. Goldscheid's rejection of Weismann drew on the energy theories of Ostwald and on Mach.
68. Rudolf Goldscheid, *Darwin als Lebenselement unserer moderne Kultur*, 1909, quoted in Hofer, "Biologen des Prater-Vivariums," 155–156. Hofer argues that Goldscheid saw Weismannism as inconsistent with Claude Bernard's milieu theory.
69. Goldscheid, *Darwin als Lebenselement*, cited in Witrisal, "Der 'Soziallamarckismus' Rudolf Goldscheids," 111.
70. See, for example, Kammerer, *Lebensbeherrschung*.
71. See Mikl-Horke, "Max Weber und Rudolf Goldscheid."
72. Kammerer, "Dualismus in der Vererbungswissenschaft."
73. Ibid., 19, and 21.
74. Kammerer, *Lebensbeherrschung*, 17
75. Kammerer, *Neuvererbung*, x.
76. Ibid., 179 and 180.

6. HEREDITY, GLANDS, AND HUMAN CONSTITUTIONS

1. Karl Sablik, *Julius Tandler, Mediziner und Sozialreformer* (Vienna: Schendl, 1983).
2. Ibid.
3. Stefan Zweig, *The World of Yesterday* [1943] (Lincoln: University of Nebraska Press, 1964), 14, 3, with permission of Viking-Penguin.
4. Steven Beller, "The World of Yesterday Revisited: Memory, and the Jews of Fin-de-Siècle Vienna," *Jewish Social Studies* 2 (1996): 37–53, 38.
5. Biographical details are from Sablik, *Julius Tandler*, 11. Tandler was not a religious man, though, like many "assimilated" Jews, he had converted to Catholicism.
6. Sablik suggests that Tandler's humanitarian leanings were clear even in his twenties, when he served in the Viennese Voluntary Rescue Society, providing emergency care to citizens.
7. Berta Zuckerkandl, *Österreich Intim: Erinnerungen 1892–1941* (Frankfurt am Main: Propyläen, 1970), 50. On medicine at the University of Vienna, see Erna Lesky, *The Vienna Medical School of the 19th Century* (Baltimore: Johns Hopkins University Press, 1976).
8. At the time Austria was often called "Deutsch-Österreich" (German-Austria).
9. *Das Neue Wien* 4 (1928): 220 (Vienna: Elbemüh, Verein für Geschichte der Arbeiterbewegung).
10. Karl Sablik, "Körner, Tandler und der 12. Februar 1934," *Mitteilungen des Instituts für Wissenschaft und Kunst* 37 (1982): 34–41; Margarete Schütte-Lihotzky, "Arzt der Gesellschaft," *Volksstimme*, August 22, 1986.
11. Julius Tandler, "Zur Histologie des äußeren Genitales," *Archiv für mikroskopische Anatomie* 54 (1899): 602–614.
12. G. Baader, "Eugenische Programme in der sozialistischen Parteienlandschaft in Deutschland und in Österreich im Vergleich," in *Eugenik in Österreich: Biopolitische Strukturen von 1900–1945*, ed. G. Baader, V. Hofer, and T. Mayer (Vienna: Czernin,

2007), 66–139. Tandler combined experimental and physiological approaches with the Austrian populism represented by Zuckerkandl; see Tatjana Buklijas, "Surgery and National Identity in Late Nineteenth-century Vienna," *Studies in the History and Philosophy of Biological and Medical Sciences* 38 (2007): 756–774.

13. Sablik, *Julius Tandler*, chapter 1; Julius Tandler, "Anatomie und Klinik," *Wiener klinische Wochenschrift* 23 (1910): 1549–1554.
14. Zuckerkandl, *Österreich Intim.*
15. Ibid., 51–52, 50.
16. For example, Julius Tandler, "Zur vergleichenden Anatomie der Kopfartierien bei den Mammalia," *Anatomische Hefte* 18 (1902): 327–363. Tandler anticipated Frank Lillie on freemartins, female cattle that are born masculinized by their brothers' sex hormones during prenatal development.
17. Chandak Sengoopta, *Otto Weininger: Sex, Science, and Self in Imperial Vienna* (Chicago: University of Chicago Press, 2000).
18. See David Luft, *Eros and Inwardness in Vienna: Weininger, Musil, Doderer* (Chicago: University of Chicago Press, 2003); and David Luft, "Thinking about Sexuality and Gender in Vienna," in *Sexuality in Austria*, ed. G. Bischof, A. Pelinka, and D. Herzog (New Brunswick, N.J.: Transaction, 2006), 21–30.
19. Julius Tandler, "Über Wachstumsstörungen," *Wiener medizinische Wochenschrift* 72 (1922): 1277–1284.
20. Ibid., 1277. Quoted with permission of Springer-Verlag Wien.
21. Ibid., 1278.
22. Ibid.
23. Ibid., 1280. Tandler especially noted interactions among the pituitary and the gonads, and the pituitary and the thyroid.
24. Julius Tandler, "Über den Einfluß der Kastration auf den Organismus, I–IV," 1909–1910, published in vols. 29–31 of *Archiv für Entwicklungsmechanik* (with S. Grosz and with K. Keller); Julius Tandler and Siegfried Grosz, *Die biologischen Grundlagen der sekundären Geschlechtscharaktere* (Berlin: Springer, 1913).
25. Tandler, "Über Wachstumsstörungen," 1280. Quoted with permission of Springer-Verlag Wien.
26. Ibid., 1283.
27. Ibid., 1279.
28. Ibid., 1284.
29. See, for example, Sarah Tracy, "George Draper and American Constitutional Medicine, 1916–1946: Reinventing the Sick Man," *Bulletin of the History of Medicine* 66 (1992): 53–89; and Carsten Timmermann, "Constitutional Medicine, Neoromanticism, and the Politics of Antimechanism in Interwar Germany," *Bulletin of the History of Medicine* 75 (2001): 717–739.
30. Julius Tandler, "Ueber den Einfluß der innersekretorischen Anteile der Geschlechtsdrüsen auf die äußere Erscheinung des Menschen," *Wiener klinische Wochenschrift* 23 (1910): 459–467. He cited skeletal changes, fat deposition before and after puberty, skin, height, and temperament as manifestations of sex gland secretions working synergistically with other glands.
31. Julius Tandler, "Konstitution und Rassenhygiene," *Zeitschrift für angewandte Anatomie und Konstitutionslehre* 1 (1913–14): 11–26. See also Sablik, *Julius Tandler,* 71–75.
32. Timmermann, "Neoromaniticism and Antimechanism." For a comparison of constitutional medicine in German and Austrian eugenics, see Baader, "Eugenische Programme."

33. Tandler, "Konstitution und Rassenhygiene," 13.
34. Ibid., 21.
35. Ibid., 22.
36. Ibid., 25.
37. Ibid., 24.
38. Letter from the Austrian Academy of Sciences to Julius Tandler, November 30, 1906. Julius Tandler Correspondence, Department and Collections, IGM.
39. Tandler, "Konstitution und Rassenhygiene," 24.
40. Von Gruber left Vienna for Munich in 1902; see Baader, "Eugenische Programme"; Paul Weindling, *Health, Race and German Politics between National Unification and Nazism 1870–1945* [1989] (Cambridge: Cambridge University Press, 1993), chapter 5. Baader writes that Gruber hoped the invitation would strengthen the relationship between the German Society and Viennese scientists. Tandler was offered membership in the German Society for Racial Hygiene, but did not accept; see Gustav Hofmann, Brigitte Kepplinger, Gerhart Marckhgott, and Hartmut Reese, *Gutachten zur Frage des Amtes der Oö. Landesregierung, 'ob der Namensgeber der Landes-Nervenklinik [Julius Wagner-Jauregg] als historisch belastet angesehen werden muss'* (Linz: Commissioned by the Amt der Oö, Landesregierung, 2005).
41. Tandler, "Konstitution und Rassenhygiene," 25.
42. Ibid., 25.
43. Ibid., 26.
44. Steinach and Kammerer used the terms "constitution" and "condition" in "Climate and Puberty."
45. Weindling, *Health, Race and German Politics*, 140.
46. Baader, "Eugenische Programme," 112–113.
47. Eugen Steinach, "Pubertätsdrüsen und Zwitterbildung," *Archiv für Entwicklungsmechanik* 42 (1916): 307–332. The figure is Table XX, Fig. 5.
48. Julius Tandler to Eugen Steinach, March 31, 1917. Julius Tandler Correspondence, Department and Collections, IGM.
49. Marianne Stein and Edmund Herrmann, "Über künstliche Entwicklungshemmung männlicher sekundärer Geschlechtsmerkmale," *Archiv für Entwicklungsmechanik* 48 (1921): 447–488.
50. Ibid., 477–478. Stein, Tandler's assistant, was one of only a few women working in the Vienna Medical School in the teen years. Tandler was considered a champion of the equal rights of women; see Joseph Friedjung, "Memorium," *Marriage Hygiene* (Bombay), November 1936, 139–141.
51. Helmut Gruber, *Red Vienna: Experiment in Working Class Culture* (Oxford: Oxford University Press, 1991). On Austromarxism, see John W. Boyer, *Culture and Political Crisis in Vienna: Christian Socialism in Power* (Chicago: University of Chicago Press, 1995); Anson Rabinbach, *The Crisis of Austrian Socialism: From Red Vienna to Civil War 1927–1934* (Chicago: University of Chicago Press, 1983); and Friedrich Stadler, "Spätaufklärung und Sozialdemokratie in Wien, 1918–1938," in *Aufbruch und Untergang: Österreichische Kultur zwischen 1918 und 1938*, ed. F. Kadrnoska (Vienna: Europaverlag, 1981), 441–473.
52. Rabinbach, *Crisis of Austrian Socialism*, 2.
53. Gruber, *Red Vienna*; Wolfgang Maderthaner and Lutz Musner, *Unruly Masses: The Other Side of Fin-de-Siècle Vienna* (New York: Berghahn Books, 2008).
54. Hofmann, Kepplinger, Marckhgott, and Reese, *Gutachten Wagner-Jauregg*. Tandler, then in exile, was not involved.

55. Friedrich Martius to Julius Tandler, April 28, 1918. Julius Tandler Correspondence, Department and Collections, IGM.
56. Veronika Hofer, "Positionen and Posen eines Experten: Die Konstitutionsforscher Julius Bauer (1887–1973) und die Eugenik in der Wiener Zwischenkriegszeit," in *Eugenik in Österreich: Biopolitische Strukturen von 1900–1945*, ed. G. Baader, V. Hofer, and T. Mayer (Vienna: Czernin Verlag, 2007), 31–65.
57. Quoted in ibid., 58.
58. Julius Bauer, *Die Konstitutionelle Disposition zu innere Krankeiten* (Berlin: Springer, 1924), 17–19, 18.
59. After 1933, Bauer challenged the racist base of the Nazi medical establishment; see Hofer, "Positionen und Posen eines Experten."
60. Jonathan Harwood, *Styles of Scientific Thought: The German Genetics Community, 1900–1933* (Chicago: University of Chicago Press, 1993), chapter 3.
61. Julius Tandler, "Konstitution und Rassenhygiene," *Mitteilungen der Volksgesundheitsamtes im Bundesministerium für Sozialverwaltung* (1920): 5–16.

7. TANDLER'S EUGENIC ENIGMAS

1. Doris Byer, *Rassenhygiene und Wohlfahrtspflege: Zur Entstehung eines sozialdemokratischen Machtdispositivs in Österreich bis 1934* (Frankfurt: Campus, 1988), 39.
2. G. Baader, "Eugenische Programme in der sozialistischen Parteienlandschaft in Deutschland und in Österreich im Vergleich," in *Eugenik in Österreich: Biopolitische Strukturen von 1900–1945*, ed. G. Baader, V. Hofer, and T. Mayer (Vienna: Czernin, 2007), 66–139; Monika Löscher, "Zur Umsetzung und Verbreitung von eugenischem/rassenhygienischem Gedankengut in Österreich bis 1924 unter besonderer Berücksichtigen Wiens," in *Medizin im Nationalsozialismus—Wege der Aufarbeitung*, ed. S. Horn and P. Malina (Vienna: ÖAW Verlag, 2001), 99–127, 111; Löscher, "Zur Popularisierung von Eugenik und Rassenhygiene in Wien," in *Wissenschaft, Politik und Öffentlichkeit*, ed. M. Ash and C. Stifter (Vienna: Facultas Verlags und Buchhandels AG, 2002), 233–265; Löscher, "Katholizismus und Eugenik in Österreich: ' . . . dass die katholische Auffassung alle vernünftigen Versuche der positive Eugenik voll Freude begrußt und unterstützt . . . ,'" in *Eugenik in Österreich: Biopolitische Strukturen von 1900–1945*, ed. G. Baader, V. Hofer, and T. Mayer (Vienna: Czernin, 2007), 140–161. See also Karin Lehner, *Verpönte Eingriffe: Sozialdemokratische Reformbestrebungen zu den Abtreibungsbestimmungen in der Zwischenkriegszeit* (Vienna: Picus, 1989); Britta McEwen, "Welfare and Eugenics: Julius Tandler's *Rassenhygienische* Vision for Interwar Vienna," *Austrian History Yearbook* 41 (2010): 170–190; and Maria Wolf, *Eugenische Vernunft: Eingriffe in die reproduktive Kultur durch die Medizin. 1900–2000* (Vienna: Böhlau, 2008).
3. Helmut Gruber, *Red Vienna: Experiment in Working Class Culture* (Oxford: Oxford University Press, 1991), chapter 2. A trial verdict that exonerated right-wing activists in the murder of two socialists triggered the 1927 revolt.
4. Carl E. Schorske, *Fin-de-Siècle Vienna: Politics and Culture* (New York: Vintage Books, 1981). On revisions of Schorske's thesis, see Steven Beller, "Introduction," in *Rethinking Vienna 1900: Austrian History Culture and Society*, ed. S. Beller (New York: Berghahn Books, 2001), 1–25.
5. John W. Boyer, *Culture and Political Crisis in Vienna: Christian Socialism in Power* (Chicago: University of Chicago Press, 1995). I rely heavily on Boyer for my account of the political situation before the war.

6. The 1905 phase of the Communist Revolution added to fears over the electoral power of the masses; ibid.
7. Gruber, *Red Vienna*, 13.
8. Bruce Pauley, "The Social and Economic Background of Austria's *Lebensunfähigkeit*," in *The Austrian Socialist Experiment: Social Democracy and Austromarxism, 1918–1934*, ed. A. Rabinbach (Boulder, Colo.: Westview, 1985), 21–37. Quote is from M. Bullock, *Austria, 1918–1938: A Story in Failure*, cited in Pauley, 23.
9. Boyer, *Culture and Political Crisis*; Maureen Healy, *Vienna and the Fall of the Habsburg Empire: Total War and Everyday Life in World War I* (Cambridge: Cambridge University Press, 2004); Marsha Rozenblit, *Reconstructing a National Identity: The Jews of Habsburg Austria during World War I* (Oxford: Oxford University Press, 2001).
10. Nancy Cartwright, Thomas Uebel, Jordi Cat, and Lola Fleck, *Otto Neurath: Philosophy between Science and Politics* (Cambridge: Cambridge University Press, 1996), 59.
11. Cited in Boyer, *Culture and Political Crisis*, 210.
12. On Austromarxism, see Gruber, *Red Vienna*, chapter 2; Rabinbach, *Crisis of Austrian Socialism*; and Friedrich Stadler, "Spätaufklärung und Sozialdemokratie in Wien, 1918–1938," in *Aufbruch und Untergang: Österreichische Kultur zwischen 1918 und 1938*, ed. F. Kadrnoska (Vienna: Europaverlag, 1981), 441–473. The Ringstrasse is Vienna's beautiful central boulevard. In the 1870s, its buildings were considered a triumphant expression of bourgeois liberal rationality.
13. *Die Arbeiter Zeitung* (Vienna), February 16, 1929.
14. Gruber, *Red Vienna*, 146.
15. Reinhard Sieder, "Housing Policy, Social Welfare, and Family Life in 'Red Vienna,' 1919–34," *Oral History* 13 (1985): 35–48.
16. Gruber, *Red Vienna*, 158 and 157.
17. Alfred Goetzl and Ralph Reynolds, *Julius Tandler: A Biography* (San Francisco: Privately printed, 1944); Gruber, *Red Vienna*; Karl Sablik, *Julius Tandler, Mediziner und Sozialreformer* (Vienna: Schendl, 1983); and Karl Sablik, "Social Politics and Health—Germany and Austria at the Time of the Weimar Republik," *Zeitschrift für die Gesamte Hygiene und ihre Grenzgebiete* 35 (1989): 225–227.
18. Lehner, *Verpönte Eingriffe;* Goetzl and Reynolds, *Julius Tandler*, 33.
19. See Michael Schwartz, "'Proletariar' und 'Lumpen': Sozialistische Ursprünge eugenischen Denkens," *Vierteljahrshefte für Zeitgeschichte* 42 (1994): 537–570.
20. Wolfgang Maderthaner and Lutz Musner, *Unruly Masses: The Other Side of Fin-de-Siècle Vienna* (New York: Berghahn Books, 2008), 17–18.
21. Gruber, *Red Vienna*; Bruce Pauley, *From Prejudice to Persecution: A History of Austrian Anti-semitism* (Chapel Hill: University of North Carolina Press, 1992); Anson Rabinbach, ed., *The Austrian Socialist Experiment: Social Democracy and Austromarxism, 1918–1934* (Boulder, Colo.: Westview, 1985).
22. Lehner, *Verpönte Eingriffe.* Lehner shows that though Goldscheid supported women's rights, the inheritance of acquired characteristics could place unusual burdens on women. See also Sieder, "Social Welfare and Family Life."
23. Gudrun Exner, "Eugenisches Gedankengut im bevölkerungswissenschaftlichen und bevölkerungspolitischen Diskurs in Österreich in der Zwischenkriegszeit," in *Eugenik in Österreich: Biopolitische Strukturen von 1900–1945*, ed. G. Baader, V. Hofer, and T. Mayer (Vienna: Czernin, 2007), 184–205; and Gudrun Exner, Josef Kytir, and Alexander Pinwinkler, *Bevölkerungswissenschaft in Österreich in der Zwischenkriegszeit (1918–1938)* (Vienna: Böhlau, 2004). Though the term *Bevölkerungspolitik* is literally translated

"population politics," Tandler's usage refers to the application of statistical demography to social problems. Many different ideologies can be implied; to avoid confusing them, I use the German term.

24. Julius Tandler, *Bevölkerungspolitische Probleme und Ziele* (Jena: Berliner medizinischen Gesellschaft, 1918), 94–110, 98–99. Reprint, Julius Tandler Papers, Department and Collections, IGM.
25. Julius Tandler, "Die wissenschaftliche Methode in sozialer Wohlfahrtsarbeit," *Österreichische Blätter für Krankenpflege* 10 (1929): 4. Reprint, Julius Tandler Papers, Department and Collections, IGM.
26. Julius Tandler, "Krieg und Bevölkerung," *Wiener klinische Wochenschrift* 29 (1916): 445–452, 9 and 20 in reprint. Emphasis added.
27. Tandler, *Bevölkerungspolitische Probleme*, 103 and 102.
28. Ibid., 105–106.
29. Tandler, "Die wissenschaftliche Methode," 2.
30. Tandler, *Bevölkerungspolitische Probleme*, 103. See also Julius Tandler, "Ehe und Bevölkerungspolitik," *Wiener medizinische Wochenschrift* 6 (1924): 211–214.
31. Tandler, "Krieg und Bevölkerung," 20.
32. Tandler, *Bevölkerungspolitische Probleme*, 102.
33. Ibid.; see also Tandler, "Ehe und Bevölkerungspolitik."
34. Tandler, *Bevölkerungspolitische Probleme*, 106.
35. Julius Tandler, "Mutterschaftszwang und Bevölkerungspolitik," *Österreichische Richterzeitung* 19 (1926): 25–29. The event celebrated Karl Kautsky's seventieth birthday; Tandler was speaking to an audience of socialists.
36. Ibid., 26.
37. Ibid., 27.
38. Tandler, "Die wissenschaftliche Methode," 4–7. Child care was still chiefly to be organized by women. When he wrote of training relief workers to aid mothers, Tandler always used feminine grammatical endings. See also Julius Tandler, "Qualitative Bevölkerungspolitik," *Mitteilungen der österreichischen Gesellschaft für Bevölkerungspolitik*, 1927, 1–17; and Julius Tandler, "The Psychology of the Fall in the Birth-rate," in *Proceedings of the World Population Conference*, ed. Margaret Sanger (London: Arnold, 1927), 208–212.
39. Tandler, "Die wissenschaftliche Methode," 7.
40. Paul Weindling, "A City Regenerated: Eugenics, Race, and Welfare in Interwar Vienna," in *Interwar Vienna: Culture between Tradition and Modernity*, ed. D. Holmes and L. Silverman (Rochester, N.Y.: Camden House, 2009), 81–113, 87.
41. Thomas Mayer, "Familie, Rasse, und Genetik: Deutschnationale Eugenikern im Österreich in der Zwischenkriegszeit," in *Eugenik in Österreich:. Biopolitische Strukturen von 1900–1945*, ed. G. Baader, V. Hofer, and T. Mayer (Vienna: Czernin, 2007), 162–183. See also H. Gabriel and W. Neugebauer, eds., *Vorreiter der Vernichtung? Eugenik, Rassenhygiene und Euthanasie in der österreichischen Diskussion vor 1938* (Vienna: Bohlau, 2005); Löscher, "Katholizismus und Eugenik"; and Weindling, "A City Regenerated."
42. Otto Reche, "Die Bedeutung der Rassenpflege für die Zukunft unseres Volkes," *Veröffentlichungen der Wiener Gesellschaft für Rassenpflege* 1 (1938): 1–15, 10). Reprint, WBR, Vienna.
43. Ibid., 2, 3, and 10. All emphasis is Reche's in the 1938 version; his emphasis is absent in the 1926 version (see offprint, Central Catalog, Österreichische Nationalbibliothek).
44. This was the same Reche to whom Kammerer and Steinach directed their criticisms in the heat rat paper, see Chapter 4. Reche's position here helps explain Kammerer's risky turn to anthropology in the paper. By doing so, Kammerer could directly

refute Reche's rejection of the inheritance of acquired characteristics and his racial approach.

45. Rudolf Polland, "Vererbung, Konstitution und Rasse in ihren Beziehung zur Heilkunde," *Ärztliche Reform-Zeitung* 26 (1924): part 1 (#19), 181–183; part 2 (# 20), 191–194; part 3 (#21), 203–204, quotes on 181, 182. A chapter on heredity in the *Hitler Handbook for Youth* also refuted the inheritance of acquired characteristics; see Robert Proctor, *Racial Hygiene: Medicine under the Nazis* (Cambridge, Mass.: Harvard University Press, 1988). On the ethical positions presumed to underlie Nazi practices, see Richard Weikart, *Hitler's Ethic: The Nazi Pursuit of Evolutionary Progress* (Basingstoke, U.K.: Palgrave Macmillan, 2011).
46. Polland, "Vererbung, Konstitution und Rasse," part 1, 183
47. Mayer, "Familie, Rasse und Genetik"; Weindling, "A City Regenerated."
48. Weindling, "A City Regenerated," 82 and 105; Löscher, "Zur Popularisierung von Eugenik."
49. For an American view, see W. Howell, "Eugenics as Viewed by the Physiologist," in M. Aldrich, W. Carruth, C. Davenport, and nine others, *Eugenics: Twelve University Lectures* (New York: Dodd and Mead, 1914), 78–106.
50. Löscher, "Zur Umsetzung und Verbreitung"; Sablik, *Julius Tandler*. On the discontinuities between the Weimar and Nazi periods in Germany, see Atina Grossmann, *Reforming Sex: The German Movement for Birth Control and Abortion Reform, 1920–1950* (Oxford: Oxford University Press, 1995).
51. See Paul Kammerer, *The Inheritance of Acquired Characteristics* (New York: Boni and Liveright, 1924), chapter 22; and Paul Kammerer, "Experiments on *Ciona* and *Alytes*," *Nature* 112 (1923): 826–827.
52. Julius Tandler, "Gefahren der Minderwertigkeit," *Das Wiener Jugendhilfswerk, Jahrbuch* (Vienna: Verlag des Jugendamts, 1928): 1–22, 1. Reprint, Julius Tandler Papers, Department and Collections, IGM. Weindling interprets the work as a shift toward sterilization; see "City Regenerated."
53. Tandler, "Gefahren," 11, 12.
54. Ibid., 12–13. Löscher, "Umsetzung und Verbreitung," describes the Marriage Advice Centers and the Union against Forced Motherhood as eugenic reforms. See also Maria Mesner, "Educating Reasonable Lovers: Sex Counseling in Austria in the first Half of the Twentieth Century," in *Sexuality in Austria*, ed. G. Bischof, A. Pelinka, and D. Herzog (New Brunswick, N.J.: Transaction, 2009), 48–64.
55. Tandler, "Gefahren," 14, 16.
56. Julius Tandler to J. Friedjung, March 2, 1936. Friedjung Handschriftensammlung, H.I.N. 203.392, WBR.
57. Tandler, "Gefahren," 15, 8, 9, 9.
58. Ibid., 4.
59. Ibid., 17.
60. Ibid., 18.
61. Ibid.
62. Ibid., 19.
63. Ibid., 21. Peter Berz and Klaus Taschwer, "Afterword," in Arthur Koestler, *Der Krötenküsser: Der Fall des Biologen Paul Kammerer* (Vienna: Czernin, 2010), summarize several replications of Kammerer's experiment.
64. Ibid., 22.
65. Ibid., 5. On Tandler and positive eugenics, see Baader, "Eugenische Programme"; and McEwen, "Welfare and Eugenics."

66. Julius Tandler diaries, generously provided by the William Tandler family.
67. From the diaries of Julius Tandler, # 1, 6, 26/VI 1929, quoted with the permission of his grandson, William Tandler.
68. The 1920 paper omits much that was in the 1913 version, but the remainder is verbatim from the 1913 version.
69. Tandler to Kammerer, November 2, 1917. Julius Tandler Correspondence, Department and Collections, IGM.
70. Robert Wistrich, "Social Democracy, Antisemitism, and the Jews of Vienna," in *Jews, Antisemitism and Culture in Vienna*, ed. I. Oxaal, M. Pollak, and G. Botz (New York: Routledge, 1987), 111–120.

8. WORKING JEWISH IN VIENNA

1. Bruce Pauley, *From Prejudice to Persecution: A History of Austrian Anti-Semitism* (Chapel Hill: University of North Carolina Press, 1992), 18.
2. Marsha Rozenblit, *Reconstructing a National Identity: The Jews of Habsburg Austria during World War I* (New York: Oxford University Press, 2001); Steven Beller, "The World of Yesterday Revisited: Memory, and the Jews of Fin-de-siècle Vienna," *Jewish Social Studies* 2 (1996): 37–53; see also Steven Beller, ed., *Rethinking Vienna 1900: Austrian History Culture and Society* (New York: Berghahn Books, 2001); Geoffrey Cantor and Marc Swetlitz, eds., *Jewish Tradition and the Challenge of Darwinism* (Chicago: University of Chicago Press, 2006); Sander Gilman, *Freud, Race, and Gender* (Princeton: Princeton University Press, 1993); Malachi Hacohen, *Karl Popper: The Formative Years, 1902–1945* (Cambridge, Mass.: Harvard University Press, 2000); David Rechtor, *The Jews of Vienna and the First World War* (Portland: Littman Library of Jewish Civilization, 2001); and Robert Wistrich, *Laboratory for World Destruction* (Lincoln: University of Nebraska Press, 2007).
3. Pauley, *Prejudice to Persecution*, 29.
4. Mitchell Hart, "Racial Science, Social Science, and the Politics of Jewish Assimilation," *Isis* 90 (2000): 268–297; Klaus Hödl, *Pathologisierung des jüdischen Körpers: Antisemitismus, Geschlecht und Medizin im Fin de Siècle* (Vienna: Picus, 1997); and Klaus Hödl, "Medizinischer Antisemitismus oder Antisemitismus in der Medizin?" in *Antisemitismusforschung in der Wissenschaft*, ed. W. Bergman and M. Körte (Berlin: Metropol, 2004), 161–185. See also John Efron, *Medicine and German Jews: A History* (New Haven: Yale University Press, 2001); Sander Gilman, *Difference and Pathology: Stereotypes of Sexuality, Race, and Madness* (Ithaca: Cornell University Press, 1985); Mitchell Hart, *The Healthy Jew* (Cambridge: Cambridge University Press, 2007).
5. His village was one of a few in the western Alps with a Jewish community organization; see Eva Grabherr, "Letters to Hohenems: A Microhistorical Study of Jewish Acculturation in the Early Decades of Emancipation," Doctoral dissertation, University of London, 2001.
6. Wolfgang L. Reiter, "Zerstört und Vergessen: Die biologische Versuchsanstalt und ihre Wissenschaftler/innen," *Österreichische Zeitschrift für Geschichte* 10 (1999): 585–614.
7. Robert Proctor, *Racial Hygiene: Medicine under the Nazis* (Cambridge, Mass.: Harvard University Press, 1988), 36.
8. Friedrich Stadler, "Spätaufklärung und Sozialdemokratie in Wien, 1918–1938," in *Aufbruch und Untergang: Österreichische Kultur zwischen 1918 und 1938*, ed. F. Kadrnoska (Vienna: Europaverlag, 1981), 441–473; Stefan Zweig, *The World of Yesterday* [1943] (Lincoln: University of Nebraska Press, 1964), 1, with permission of Viking-Penguin.

9. Reprinted in *Sigmund Freud: His Life in Pictures*, ed. Ernst Freud, Lucie Freud, and Ilse Grubrich-Simitis (Ontario: Penguin Books, 1985), 258–259. The quote is from the poem.
10. Steinach to Harry Benjamin, June 26, 1924. Benjamin-Steinach Correspondence, Historical Collections, NYAM.
11. On Billroth, see Pauley, *Prejudice to Persecution*. For some assimilated Jews even the horrific events in the early 1940s were difficult to believe; see Peter Singer, *Pushing Time Away: My Grandfather and the Tragedy of Jewish Vienna* (London: Granta Books, 2003).
12. Malachi Haim Hacohen, "Dilemmas of Cosmopolitanism: Karl Popper, Jewish Identity, and 'Central European Culture,'" *Journal of Modern History* 71 (1999): 105–149; Fritz Ringer, *Decline of the German Mandarins: The German Academic Community, 1890–1933* [1969] (Hanover: University Press of New England, 1990).
13. Wistrich, *World Destruction*, 8–9.
14. Marsha Rozenblit, *The Jews of Vienna, 1867–1914: Assimilation and Identity* (Albany: State University of New York Press, 1983), chapter 5. Immigrant Jewish fathers sent their sons to Gymnasien more commonly than immigrant fathers of non-Jewish sons.
15. Ibid.; Rozenblit, *Reconstructing National Identity*.
16. Hacohen, "Dilemmas of Cosmopolitanism"; and Malachi Haim Hacohen, "The Culture of Viennese Science and the Riddle of Austrian Liberalism," *Modern Intellectual History* 6 (2009): 369–396; see also Gilman, *Freud, Race, and Gender;* Alan Janik, "Vienna 1900 Revisited," in *Rethinking Vienna 1900*, ed. S. Beller (New York: Berghahn, 2001), 27–56; and Wistrich, *World Destruction*. On nineteenth-century German nationalism, see David Sorkin, *The Triumph of German Jewry, 1780–1840* (New York: Oxford University Press, 1987); John Weiss, *Ideology of Death: Why the Holocaust Happened in Germany* (Chicago: Ivan Dee, 1997).
17. Hacohen, "Dilemmas of Cosmopolitanism," 116 (paraphrasing Adolf Jellinik). Zionism was an exception; see Wistrich, *World Destruction;* and Mitchell Hart, *Social Science and the Politics of Modern Jewish Identity* (Palo Alto: Stanford University Press, 2000).
18. Hacohen, "Dilemmas of Cosmopolitanism," 106; see also Beller, *Rethinking Vienna*.
19. Hart, "Racial Science and Jewish Assimilation," 269. Hart shows that the environmentalist stance of several Jewish thinkers was also used to explain the degeneration of the "race." See also Veronika Lipphardt, *Biologie der Juden: Jüdische Wissenschaftler über "Rasse" und Vererbung, 1900–1935* (Göttingen: Vandenhoeck and Ruprecht, 2008).
20. Hacohen, "Dilemmas of Cosmopolitanism," 180.
21. Klaus Hödl, "Julius Tandler und das 'Rote Wien,'" *Medizin und Judentum* 5 (2000): 112–120, 112; and Hödl, *Pathologisierung des jüdischen Körpers;* Karl Sablik, *Julius Tandler, Mediziner und Sozialreformer* (Vienna: Schendl, 1983), chapter 1, 31. Tandler also challenged Reche's position on the importance of hiring Aryans at the university; see Lipphardt. *Biologie der Juden*, 225.
22. Sigmund Freud, *The Diary of Sigmund Freud, 1929–1939*. Translated and annotated by Michael Molner (New York: Scribners, 1992), 45; Sablik, *Julius Tandler*, chapter 4.
23. Sablik, *Julius Tandler*.
24. Hödl, "Tandler und 'das Rote Wien,'" 113.
25. On Sigmund Freud's reaction to antisemitic critiques of psychoanalysis, see Sander Gilman, *The Case of Sigmund Freud: Medicine and Identity at the Fin de Siècle* (Baltimore: Johns Hopkins University Press, 1993).
26. Hödl, "Tandler und "das Rote Wien," 114; from the diaries of Julius Tandler, #5.2, 3, Montag 11./IX, quoted with the permission of his grandson, William Tandler; Lipphardt, *Biologie der Juden*.

27. Klaus Hödl, "Jüdischer Identitätswandel im Kontext der Rassenhygiene," *Geschichte und Gegenwart* 3 (1999): 131–146.
28. Sablik, *Julius Tandler*, chapter 4.
29. Letter from Ernst Ress to Julius Tandler, October 31, 1931; Julius Tandler Correspondence, Department and Collections, IGM.
30. Sablik, *Julius Tandler*.
31. Tandler rejected Jewish nationalism as no different from any other; see Tandler Papers, Department and Collections, IGM.
32. On the Wiederspergs, see Arthur Koestler, *The Case of the Midwife Toad* (New York: Random House, 1971); Protocols of Vivarium meetings, 1921–27, Vivarium Archives, ÖAW. I thank Sabine Brauckmann for this information. See Kammerer to H. Glücksmann, April 16, 1920, Kammerer Correspondence, ÖNB.
33. His last hope was a professorship from the Communist Academy in Moscow in 1926. When he died, Kammerer was preparing to set up a laboratory there; Peter Berz and Klaus Taschwer, "Afterword," in Arthur Koestler, *Der Krötenküsser: Der Fall des Biologen Paul Kammerer* (Vienna: Czernin, 2010).
34. Kammerer to Benjamin, July 11, 1923, and August 31, 1923, Harry Benjamin Papers, Correspondence, Paul Kammerer Folder 2, KI. Berz and Taschwer note the contrast between Kammerer's happy young adulthood and his life in the 1920s; see "Afterword."
35. Albrecht Hirschmüller, "Paul Kammerer und die Vererbung erworbener Eigenschaften," *Medizinhistorisches Journal* 26 (1991): 26–77, 46.
36. T. H. Morgan to Paul Maerker-Branden, October 1, 1923, William Bateson Papers, Microfilm #26, APS. Quoted with the permission of the Syndics of Cambridge University Library.
37. *New York Times*, December 2, 1923; November 29, 1923; December 9, 1923; November 28, 1923; November 25, 1923. An article on his suicide, probably by Benjamin, indicated that critics stressed Kammerer's socialism and his efforts to popularize science; see *New York Times*, September 26, 1926.
38. Hans Przibram, "Paul Kammerer als Biologe, *Monistische Monatshefte* 11 (1926): 401–405, 405.
39. Przibram had also expressed this view well before the fraud was discovered, see Hans Przibram to Hugo Iltis, July 23, 1923, Kammerer Papers, Iltis Correspondence, APS.
40. Berz and Taschwer, "Afterword," 312–313; Hirschmüller, "Paul Kammerer"; "Der Selbstmord Paul Kammerers," *Neue Freie Presse*, September 25, 1926.
41. Paul Kammerer, *Menschheitswende: Wanderungen im Grenzgebiet von Politik und Wissenschaft* (Vienna: Der Friede, 1919).
42. Sander Gliboff, "The Pebble and the Planet: Paul Kammerer, Ernst Haeckel, and the Meaning of Darwinism," Ph.D. dissertation, Johns Hopkins University, 2001, 186. Gliboff also describes antisemitic slurs in correspondence addressed to August Weismann (see 187–188).
43. Pearl to Donaldson, December 20, 1923. Raymond Pearl Papers, APS. On Pearl's antisemitism, see Elazar Barkan, *The Retreat of Scientific Racism: Changing Concepts of Race in Britain and the United States between the World Wars* (Cambridge: Cambridge University Press, 1993).
44. Erwin Baur, Eugen Fischer, and Fritz Lenz, *Grundriss der menschlichen Erblichkeitslehre und Rassenhygiene*, volumes 1–2 (Berlin: Lehmann, 1921).
45. Proctor, *Racial Hygiene*, chapter 2; Sheila Weiss, "Race and Class in Fritz Lenz's Eugenics," *Medizinhistorisches Journal 27* (1992): 5–25.

46. Proctor, quoting Lenz, *Racial Hygiene*, 55. The mention of writing about Goethe probably referred to Kammerer, who had done so.
47. Fritz Lenz to Hugo Iltis, July 25, 1924. Hugo Iltis Collection, Special Collections, UWL. Lenz referred to an unnamed American scientist who also believed that the inheritance of acquired characteristics was a Jewish racial trait.
48. Fritz Lenz, "Der Fall Kammerer und seine Umfilmung durch Lunatscharsky," *Archiv für Rassen- und Gesellschafts-biologie* 21 (1929): 311–318, 311, 316.
49. Hugo Iltis to his son Hugh Iltis, 1951, typescript, Kammerer Papers, APS.
50. Julius Tandler, "Konstitution und Rassenhygiene," *Zeitschrift für angewandte Anatomie und Konstitutionslehre* 1 (1913–14): 11–26, 20.
51. Hart, "Racial Science and Jewish Assimilation."
52. Proctor, *Racial Hygiene*, 36, 30. Lang cited Weismann.

9. ASYMMETRY, FAILURE, AND FLEXIBLE HEREDITY

1. Richard Goldschmidt, "Fifty Years of Genetics," *American Naturalist* 84 (1950): 313–339, 319, 320. Goldschmidt was a German geneticist and a Jewish émigré who fled to America in the 1930s; he often challenged aspects of genetics (see Marsha Richmond, "The Cell as the Basis for Heredity, Development, and Evolution: Richard Goldschmidt's Program of Physiological Genetics," in *From Embryology to Evo-Devo: A History of Developmental Evolution*, ed. Manfred D. Laubichler and Jane Maienschein (Cambridge, Mass.: MIT Press, 2007), 169–212.). Kammerer knew Goldschmidt's work and lamented his growing reticence about the inheritance of acquired characteristics. After the suicide, Goldschmidt became convinced that Kammerer's evidence was fraudulent; see Peter Berz and Klaus Taschwer, "Afterword," in Arthur Koestler, *Der Krötenküsser: Der Fall des Biologen Paul Kammerer* (Vienna: Czernin, 2010).
2. Snait Gissis and Eva Jablonka, "The Modern Synthesis," in *Transformations of Lamarckism: From Subtle Fluids to Molecular Biology*, ed. Snait Gissis and Eva Jablonka (Cambridge, Mass.: MIT Press, 2011), 101–142.
3. Peter Bowler, *The Eclipse of Darwinism* (Baltimore: Johns Hopkins University Press, 1983), 60; my emphasis.
4. Amid a rich literature, see, for example, Ian Hacking, "The Looping Effects of Human Kinds," in *Causal Cognition*, ed. D. Sperber, D. Premack, and A. Premack (New York: Oxford University Press, 1995), 351–383; Ian Hacking, *Rewriting the Soul: Multiple Personality and the Sciences of Memory* (Princeton: Princeton University Press, 1995); Bruno Latour; *Science in Action: How to Follow Scientists and Engineers through Society* (Cambridge, Mass.: Harvard University Press, 1986); and Hans-Jörg Rheinberger and Michael Hagner, eds., *Die Experimentalisierung des Lebens: Experimentalsysteme in den biologischen Wissenschaften 1850/1950* (Berlin: Akademie-Verlag, 1993).
5. In 1935, American geneticist A. Franklin Shull mused that adopting the language of mathematics had been most effective in speeding up the neo-Darwinist view that evolution was a matter of gene frequencies: "How our friends the economists and psychologists must envy us now in the possession of one small means of reducing the number of those who feel qualified to speak." Shull, "Weismann and Haeckel: One Hundred Years," *Science* 81 (1935): 443–452, 448.
6. Bruno Latour presents the subtlety of the process eloquently in *Aramis, or the Love of Technology* (Cambridge, Mass.: Harvard University Press, 1996). I thank Sarah Wagner for suggesting the relevance of Latour's example.

7. Robert Kohler, *Lords of the Fly: Drosophila Genetics and the Experimental Life* (Chicago: University of Chicago Press, 1994).
8. Latour, *Aramis*, 10.
9. Heinrich E. Ziegler, "Die Streitfrage der Vererbungslehre (Lamarckismus oder Weismannismus)," *Naturwissenschaftliche Wochenschrift* 9 (1910): 194–202. On Ziegler's biology and politics, see Paul Weindling, *Health, Race and German Politics between National Unification and Nazism 1870–1945* [1989] (Cambridge: Cambridge University Press, 1993), chapter 2.
10. Ziegler's rejection of Lamarckism altered the meaning of "instinct," basing the concept in the selection of inherited neural pathways; see Ziegler, "Streitfrage,"13; on Ziegler and instinct, see Richard Burkhardt, Jr., *Patterns of Behavior: Konrad Lorenz, Niko Tinbergen, and the Founding of Ethology* (Chicago: University of Chicago Press, 2005). Ziegler's rejection is also based in a different interpretation of the gonad than that of Tandler, Steinach, and Kammerer. The difference illustrates the significance of their assertion that the gonad is a double gland.
11. Paul Kammerer, *Neuvererbung oder Vererbung erworbener Eigenschaften: Erbliche Belastung und erbliche Entlastung* (Stuttgart: Walter Seifert, 1925), 174.
12. Ziegler helped frame the announcement of the Krupp Prize, which popularized eugenics in Germany; see Weindling, *Heath, Race and German Politics.*
13. Henry Cowles, "The Adaptation Theory," *Botanical Gazette* 38 (1904): 385–386, 386.
14. Robert Proctor, *Racial Hygiene: Medicine under the Nazis* (Cambridge, Mass.: Harvard University Press, 1988), 297.

EPILOGUE

1. See Denis Thieffry and Sahotra Sarkar, "Forty Years under the Central Dogma," *Trends in Biochemical Sciences* 23 (1998): 312–316; and Snait Gissis and Eva Jablonka, eds., *Transformations of Lamarckism: From Subtle Fluids to Molecular Biology* (Cambridge, Mass.: MIT Press, 2011).
2. See, for example, Frances A. Champagne, "Epigenetic Mechanisms and the Transgenerational Effects of Maternal Care," *Frontiers in Neuroendocrinology* 29 (2008): 386–397; D. C. Dolinoy, D. Huang, and R. L. Jirtle, "Maternal Nutrient Supplement Counteracts Bisphenol a-induced DNA Hypomethylation in Early Development," *Proceedings of the National Academy of Sciences* 104 (2007): 13056–13061; P. D. Gluckman, Mark Hanson, and Alan Beedle, "Non-genomic Transgenerational Inheritance of Disease Risk," *BioEssays* 29 (2007): 145–154; Gilbert Gottlieb, *Synthesizing Nature and Nurture: Prenatal Roots of Instinctive Behavior* (Mahwah, N.J.: Erlbaum, 1997); Susan Oyama, Paul E. Griffiths, and Russell D. Gray, eds., *Cycles of Contingency: Developmental Systems and Evolution* (Cambridge, Mass.: MIT Press, 2001); C. Wolf and D.E.J. Linden, "Biological Pathways to Adaptability—Interactions between Genome, Epigenome, Nervous System, and Environment for Adaptive Behavior," *Genes, Brain and Behavior* 11 (2012): 3–28.
3. Gissis and Jablonka, *Transformations of Lamarckism.*
4. See Sander Gliboff, "Did Paul Kammerer Discover Epigenetic Inheritance? No and Why Not," *Journal of Experimental Zoology, B Molecular Development and Evolution* 314B (2010): 616–624; and A. O. Vargas, "Did Paul Kammerer Discover Epigenetic Inheritance? A Modern Look at the Controversial Midwife Toad Experiments," *Journal of Experimental Zoology, B. Molecular Development and Evolution* 312 (2009): 667–678.

INDEX

Note: Figures are indicated by the letter *f.* Notes are indicated by the letter *n* inserted after the page number and before the note number.

ABOUT THE AUTHOR

CHERYL LOGAN was trained as an experimental psychologist and for several years conducted field research on wild songbirds. She now pursues a long-standing interest in the history of biology and the life sciences as an Emerita Professor of Psychology and History at the University of North Carolina Greensboro.